1980 J.C.T. Standard Form of Building Contract

A Commentary

Macmillan Building and Surveying Series
Series Editor: Ivor H. Seeley
Emeritus Professor, Nottingham Polytechnic

Advanced Building Measurement, second edition Ivor H. Seeley
Advanced Valuation Diane Butler and David Richmond
An Introduction to Building Services Christopher A. Howard
Applied Valuation Diane Butler
Asset Valuation Michael Rayner
Building Economics, third edition Ivor H. Seeley
Building Maintenance, second edition Ivor H. Seeley
Building Procurement Alan Turner
Building Quantities Explained, fourth edition Ivor H. Seeley
Building Surveys, Reports and Dilapidations Ivor H. Seeley
Building Technology, third edition Ivor H. Seeley
Civil Engineering Contract Administration and Control Ivor H. Seeley
Civil Engineering Quantities, fourth edition Ivor H. Seeley
Civil Engineering Specification, second edition Ivor H. Seeley
Computers and Quantity Surveyors Adrian Smith
Contract Planning and Contractual Procedures B. Cooke
Contract Planning Case Studies B. Cooke
Environmental Science in Building, second edition R. McMullan
Housing Associations Helen Cope
Introduction to Valuation D. Richmond
Principles of Property Investment and Pricing W.D. Fraser
Quality Assurance in Building Alan Griffith
Quantity Surveying Practice Ivor H. Seeley
Structural Detailing P. Newton
Urban Land Economics and Public Policy, fourth edition P.N. Balchin, J.L. Kieve and G.H. Bull
Urban Renewal Chris Couch
1980 JCT Standard Form of Building Contract, second edition R.F. Fellows

Other titles by the same authors
Housing Improvement and Social Inequality P.N. Balchin (Gower)
Housing Policy: An Introduction P.N. Balchin (Croom Helm)
Housing Policy and Housing Needs P.N. Balchin (Macmillan)
Regional and Urban Economics P.N. Balchin and G.H. Bull (Harper and Row)
The Electric Telegraph: An Economic and Social History J.L. Kieve (David and Charles)

1980 J.C.T. Standard Form of Building Contract

A Commentary for Students and Practitioners

R. F. Fellows

Construction Study Unit

School of Architecture and Building Engineering

University of Bath

Second Edition

First edition 1981
Reprinted 1982, 1983, 1985
Second Edition 1988
Reprinted 1990 (twice)

Published by
MACMILLAN EDUCATION LTD
Houndmills, Basingstoke, Hampshire RG21 2XS
and London
Companies and representatives
throughout the world

Printed in Hong Kong

British Library Cataloguing in Publication Data
Fellows, R.F.
1980 JCT standard form of building contract:
a commentary for students and practitioners.
—2nd ed. — (Macmillan building and
surveying).
1. Joint Contracts Tribunal. Standard form
of building contract. 1980 2. Building
—Contracts and specifications—
Great Britain
I. Title
692'.8
TH425

ISBN 0–333–46325–0

Contents

Preface

This book is designed to provide an introduction to the complex contractual situations encountered in the building industry, directly associated with the use of the J.C.T. Standard Form of Building Contract, 1980 Edition.

In 1980 the Joint Contracts Tribunal issued a new set of standard contracts for use on building projects. These documents contain some extensive modifications to the standard contracts (also mainly of J.C.T. origin) which were in wide use. The new contracts not only modified the previous editions but also, by express terms, changed the applicability of some case precedents which had become well known and widely recognised parameters for building operations.

The objective of a standard contract is to provide a clear and unambiguous contract in order to prevent disputes arising out of the terms of the contract and interpretation thereof. Note that provisions in the Unfair Contract Terms Act, 1977, refer to standard form contracts and to consumer contracts; JCT '80 is of the former type.

The legal knowledge required by anyone involved with the use and interpretation of contracts changes and increases daily, so constant attention to the evolution of, particularly, statute and case law is essential.

The intention is to provide an interpretation of and commentary upon the terms of the J.C.T. Standard Form of Building Contract, Private with Quantities, 1980 Edition, combined with considerations of the relevant precedents. Therefore, the aim is to provide a guide for 'every day' use. In order to pursue any aspect of the contract in great detail, the reader is advised to consult one of the available definitive authorities (such as Hudson's Building and Engineering Contracts).

As this book is intended to serve many purposes, the reader requires no previous knowledge of the standard contracts used in the building industry but will find it of value to have an appreciation of basic English law, particularly that relating to contracts.

The second edition of this book incorporates the amendments to JCT '80 published between its launch and February 1987: these consider ownership of materials on site, NSC/1A and the indemnity and insurance provisions. Recent development in case law have been included.

Thus, this book will prove of use to many concerned with building, whether in industry or the professions commonly encountering problems of interpretation and implementation of the contract, or as students.

The use of this book is recommended to be in conjunction with a copy of the appropriate J.C.T. contract in order that the exact terminology of the document may be studied together with its interpretation. This is particularly important in practical situations where amendments to the contract vary the standard terms.

Acknowledgements

My thanks are due to many people who have been of help in the preparation of this book.

In particular, I am grateful to Dr. I. H. Seeley for his most useful comments after reading a full draft, to David Langford and Robert Newcombe for their very objective comments and encouragement during the period of writing and to Mr. Malcolm Stewart for his patience and care in the publication.

Finally, my thanks to Marianne Bevis for her incomparable tolerance and diligence in typing the final manuscript and, of course, to my wife, Shirley McCarthy, for her unceasing support and assistance.

Richard Fellows
Uxbridge, January 1981

2nd Edition:

In addition to those people who provided much help with the preparation of the original edition of this book, I am grateful for the considerable assistance rendered by the following people in the production of the second edition:

Stan Hutton,

Peter Murby for his efforts in publishing the volume, and

Lisa Warr for her patience and expertise in typing the manuscript.

Responsibility for any errors and omissions in this book remains mine alone.

Richard Fellows
Uxbridge, February 1987

Abbreviations

Q.S.	Quantity Surveyor
D.L.P.	Defects Liability Period
NS/C	Nominated Sub-Contractor
NSC	Nominated Sub-Contract
NSup	Nominated Supplier
B.Q.	Bill(s) of Quantities
S.M.M.6	Standard Method of Measurement of Building Works: Sixth Edition
N.E.D.O.	National Economic Development Office
N.J.C.B.I.	National Joint Council for the Building Industry
C.I.T.B.	Construction of Industry Training Board
C.o.W	Clerk of Works
S/C	Sub-Contractor
S.O.	Supervising Officer
V.A.T.	Value Added Tax
P.C.Sum	Prime Cost Sum
A.I.	Architect's Instruction
F.I.D.I.C.	Federation Internationale des Ingenieurs-Conseils
L.A.	Local Authority
J.C.T.	Joint Contracts Tribunal
S.I.	Statutory Instrument
S.	Section (in reference to a statute)
U.K.	United Kingdom
P.A.Y.E.	Pay as You Earn
N.I.	National Insurance
P.Q.S.	Private Quantity Surveyor
R.I.B.A.	Royal Institute of British Architects
C.D.P.	Contractor's Designed Portion

Use of this Book

This book fulfils three primary functions: as a textbook, as a source to amend and update knowledge of the 1963 J.C.T. Standard Form, as a practical aid to interpretation and use of the 1980 J.C.T. Standard Form or Building Contract.

It is recommended, therefore, that a copy of the 1980 J.C.T. Standard Form be kept available by the reader to enhance consideration of precise points of detail. In instances of industrial and professional use, reference should be made to a copy of the Contract Document used upon the project under examination (especially where amendments to the standard contract have been incorporated). Reference should also be made to all relevant law reports to ensure compliance with precedents, including recent amendments thereto.

The reader requires no prior knowledge of specialist building contracts, although a basic knowledge of English law, especially contract and tort, and of the building process will prove advantageous.

The reader may occasionally require more detailed discussion than it has been possible to include in this volume; in such circumstances reference is recommended to the works contained in the Bibliography/Sources section to be found at the back of this book.

To accord with the terminology used by the J.C.T., the style of writing employed in this book uses the masculine gender. In all cases, it is the intention of the author that masculine versions of words and terms refer equally to their feminine equivalents. No sex discrimination whatsoever is intended and it is hoped that styles of writing will change soon to eliminate such apparent inequity.

Introduction

Under this form of contract, as under its predecessors (R.I.B.A. Form, 1963 JCT contract), the contractor undertakes to carry out the work under described conditions. Details of the required work and conditions must be set out in the Contract Documents to be valid.

There are also general principles of law relating to any activity, including construction, which must be observed, e.g. public liabilities, statutes being the primary source of law which must be applied in preference to any other requirements.

Thus the Factories Acts, Building Regulations, etc., must always be adhered to even though not specifically incorporated as Contract Documents. They are incorporated into the Conditions of Contract by Clause 6.

The Contract Documents, most particularly the Conditions of Contract, set out the express terms of the contract. Usually, however, further terms must be implied in the contract to give it business efficacy.

Cory v. City of London Corporation (1951). "In general a term is necessarily implied in any contract, the other terms of which do not repel the implication, that neither party shall prevent the other from performing it."

Trollope & Colls Ltd. v. N.W. Metropolitan Regional Hospital Board (1973). "An unexpressed term can be implied if and only if the court finds that the parties must have intended that term to form part of their contract."

Contract Documents (Clauses 2 and 5)
Clause 2.1 specifies the Contract Documents to be:

(a) Contract Drawings
(b) Contract Bills
(c) Articles of Agreement
(d) Conditions of Contract
(e) Appendix to the Conditions

However, certain other documents are also incorporated, apart from statute and common law provisions. These are:

(f) S.M.M.6 (Clause 2.2.2.1)
(g) Basic Price List, if applicable (Clause 39.1)
(h) Formula Rules, Monthly Bulletins, if applicable (Clause 40)
(i) Form of Tender

Form of Tender
Should there be a discrepancy between the Form of Tender and the Conditions of Contract, etc., provided the Contract had been signed by the relevant parties, it is probable that the Contract Conditions would prevail.

Specification and Bills of Quantities
The Specification is not a contract document under this form although it is a Contract Document in the Without Quantities form.

The B.Q. therefore must carry out a multitude of functions, inter alia;

(a) an exact measure of the work to be completed for the contract sum in terms of both quality and quantity
(b) provides a basis for the measurement and valuation of Variations
(c) provides a means of incorporating the necessary specification information as part of a contract document.

Contractual Relationships
The Parties to the contract are:
The Contractor
The Client (called the Employer)

The Architect, Q.S., Engineer, C.o.W, and other consultants, are not parties to the contract. Each has their own terms of employment with the Employer (e.g. Architect - Conditions of Engagement), usually of a standard form issued by the appropriate professional institution. Often such standard terms set out fee scales - R.I.C.S. advisory; R.I.B.A. advisory.

The Architect is given quite extensive powers of Agency (of the Employer) under the JCT Standard Form in respect of the work. He may not alter the terms of the Contract itself (to which he is not a party).

Clauses 27 and 28 provide for determination of the Contrac
Employment in certain circumstances. If such determination occu
it is the employment of the Contractor under the contract which
ceases, prescribed terms of the contract remain in force to facilitate
settlement between the parties. These clauses are in addition to
Common Law rights where, if one party is in fundamental breach of a
contract, the aggrieved party is relieved of any outstanding
obligations under the contract and usually will have a right to
damages also.

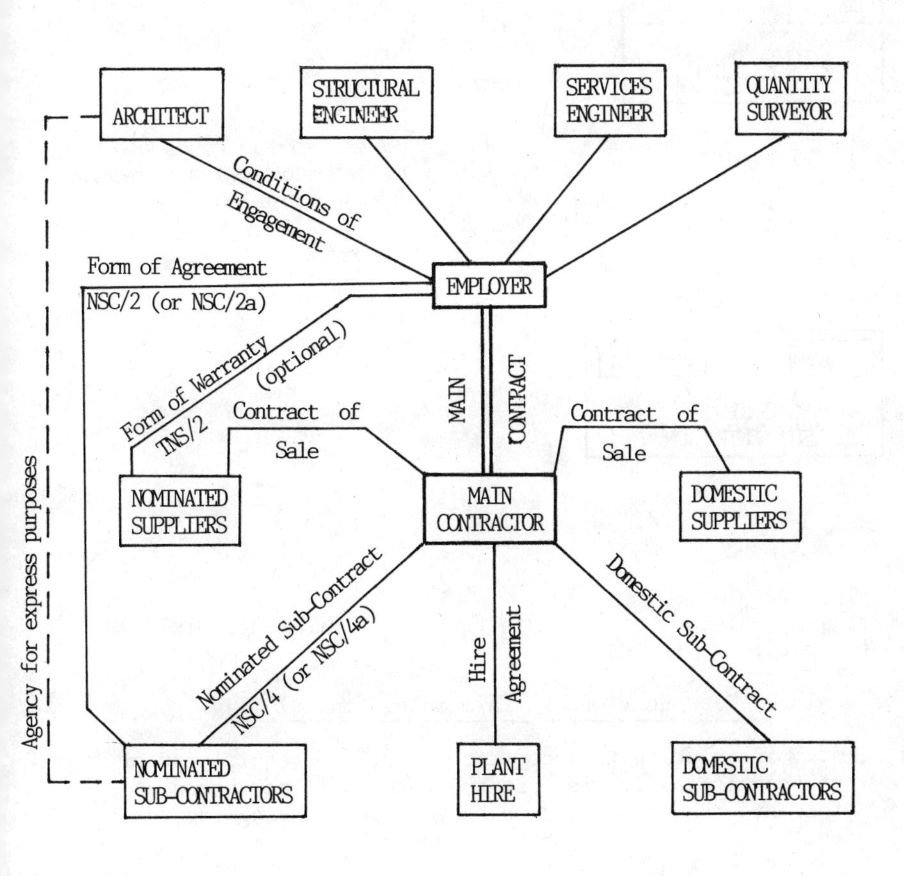

The Nomination components may be further expanded thus:

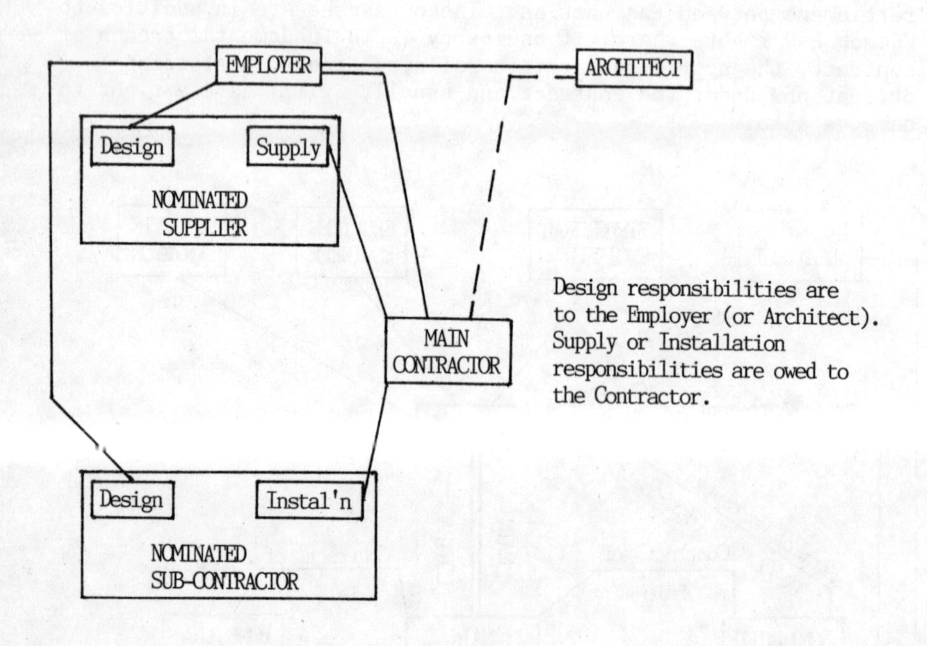

Certain provisions of the Unfair Contract Terms Act, 1977, should be noted:

(see also: McCrone v. Boots Farm Sales Ltd. (1981))

S11(1) : '... in relation to a contract term, the requirement of reasonableness ... is that the term shall have been a fair and reasonable one to be included, having regard to the circumstances which were, or ought reasonably to have been, known, or in the contemplation of the parties when the contract was made'.

S17 : 'Any term of a contract which is a consumer contract or a standard form contract shall have no effect for the purpose of enabling a party to the contract' to avoid his contractual liabilities unless it is reasonable.

ARTICLES OF AGREEMENT

The Articles of Agreement constitute the actual contract between the parties whilst the Conditions of Contract stipulate certain provisions for its execution.

Note: 50p stamp* to be affixed to the Articles of Agreement if the contract is to be under seal. By the Limitation Acts, 1939-1980, an action upon a simple contract must be brought within 6 years of the accrual of the cause of action, 12 years if upon a contract under seal. *(Requirement for stamp - abolished under the Finance Act, 1985.)

Preliminary Article
Sets out certain basics of the contract:

(a)	Date	
(b)	Parties - Employer and Contractor	
(c)	Defines and locates the Works	
(d)	Defines who has been responsible for the preparation of Drawings and B.Q.	1st Recital
(e)	Contractor has given Employer fully priced B.Q. - 'Contract Bills'	2nd Recital
(f)	Numbers the 'Contract Drawings'	3rd Recital
(g)	Status of the Employer under Finance (No.2) Act 1975 as at the Date of Tender - refer to Appendix (Relates to statutory tax deduction scheme)	4th Recital

Article 1
Over-riding obligation on the Contractor to complete the

Works, under the described conditions for the Contract Sum.

Article 2
Defines the Contract Sum and refers to the Conditions for the payment times and the payment method.

See Hoenig v. Isaacs (1952)
Lump sum Contract even though there are specific provisions for periodic interim payments.

Article 3A - Article 3 in Private Edition
Specifies the Architect
Architect - registered under Architects' Registration Acts, 1931-1969. Architect's death or named Architect ceasing to act as such - Employer must re-nominate another Architect. Contractor has right to object to new Architect, subject to arbitration provisions of Article 5. A replacing Architect cannot over-rule decisions, instructions, certificates and other similar items issued by previous Architect.

In Stanley Hugh Leach Ltd. v. L. B. Merton (1985) the judge accepted that, under JCT '63, the employer was obliged to appoint a person who would be "reasonably competent and would use that degree of diligence, skill and care in carrying out the duties assigned to him". Due to similarity of wording, this probably will apply to JCT '80.

Article 3B - L.A. Form only - not in Private Edition
To be used where the Employer will have a supervising officer instead of an Architect.

Common for Local Authorities.

Same provisions as for Article 3A.

The option of having a Supervising Officer instead of an Architect is a major difference between the Local Authority and Private Forms.

Articles 3A and B
The appointment/re-appointment of Architect etc., is a condition of the Contract - it goes to the root of things.

Note: If the formula fluctuations are applicable under Clause 40, then Clause 40.2 amends the interim valuations provisions of Clause 30.1.2 by requiring an interim valuation to be carried out prior to the issues of each Interim Certificate.

As the Q.S. is obliged to carry out interim valuations it is submitted that, where Clause 40 is used, the appointment/re-appointment of the Q.S. is a condition (rather than a warranty).

This is evidenced by such provision as the role of the Architect in issuing certificates. Non-appointment/re-appointment will be grounds for determination unless the Contract is endorsed to waive this requirement. (Barnes v. Landair Holdings Limited (1974).)

Re-appointment must always be of the same status, in the absence of a specific agreement to the contrary.

The S.O. performs the functions of an Architect under the contract but, due to lack of qualifications and registration, is not permitted to be called an Architect.

Article 4
Specifies the Q.S. Similar re-appointment provisions as for Architect.

Note: No specific provision to exclude new Q.S. reversing decisions of previous Q.S.

Appointment/re-appointment - failure to appoint/re-appoint - claim for damages for breach of contract.

Q.S. responsibilities under the contract: (see also R. B. Burden v. Swansea Corporation (1957).)

5.1	custody of Contract Drawings and Bills
13.4.1	value Variations under Clause 13.4 rules, unless otherwise agreed by Contractor and Employer
13.6	allow Contractor to be present and take notes etc., for valuing Variations
26.1 & 26.4.1	ascertain amount of Contractor's loss/expense - delays - Architect's option make Interim Valuations.
30.1.2	make Interim Valuations
30.5.2.1	statement of Retentions - Architect's option
30.6.1.2	statement of Final Valuations of Variations
34.4.1	ascertain amount of Contractor's loss/expense - antiquities - Architect's option
38.4.3	agreement with Contractor of amount of fluctuations
39.5.3	agreement with Contractor of amount of fluctuations

40.5 agree with Contractor any alterations of formula fluctuations recovery methods

Note: The Architect, the Supervising Officer and the Quantity Surveyor must carry out their functions under the Contract in the manners expressly provided by the Contract, unless the parties to the Contract agree otherwise.

Article 5 - Settlement of Disputes - Arbitration (previously Clause 35 - 1963 Edition)

5.1 Defines disputes to be referred to arbitration and the person to arbitrate:
Dispute between Architect, Employer and the Contractor
Arising during the progress, after completion or abandonment
About - construction of the Contract
- anything arising under the Contract Note: discretion of Architect.

Excluding - Statutory Tax Deduction Scheme (Clause 31.9); Possible exemption from V.A.T.

Arbitrator - agreed by the parties - no agreement within 14 days of one party's notice to request concurrence in appointment - upon request - appointed by President or Vice-President of the R.I.B.A.

.4 Related disputes of a similar nature (Employer - NS/C; Contractor - NS/C; Contractor/Employer - NSup) to be referred to that Arbitrator to decide both disputes but not
.5 if either Employer or Contractor considers the Arbitrator to be insufficiently qualified to hear the dispute.

Note: 5.1.4 and 5.1.5 are optional (may be deleted) - see 5.1.6.

5.2 Arbitration is usually after Practical Completion or determination of the Contractor's employment (both actual or alleged) or abandonment of the Works, unless commenced earlier with written consent of Employer, Architect and the Contractor.

Exceptions: re-appointment of Architect, Q.S.,
dispute re. power to issue an instruction
dispute re. improper withholding of a certificate
dispute re. certificate not being in accordance with the Conditions
dispute re. reasonable objection by Contractor to an A.I. (Clause 4.1)

dispute re. extension of time
dispute re. outbreak of hostilities
dispute re. war damage.

5.3 Arbitrators have the power to decide all matters referred and to order all necessary measurement, etc., to be made. Arbitrators may and should use any special knowledge they possess in weighing the evidence and making an award (Anne Fox v. P. G. Wellfair Ltd.).

In Northern Regional Health Authority v. Derek Crounch Construction Co. Ltd. (1983), the Court of Appeal determined: a) that there is no rule of law requiring an Arbitrator to decide all matters referred, and b) under JCT contracts, Arbitrators are given power to open up, review and revise any 'certificate, decision, opinion, decision requirement or notice' but such power does not apply to the Courts which may decide only whether a breach of contract occurred.

5.4 Arbitrator's award to be final and binding on the parties.

5.5 Unless amended; in all cases law of England applies to all Arbitrations under Arbitration Acts, 1950 and 1979.

This provision will normally be left unamended as England is generally recognised as being the world's centre for arbitrations - this is recognised by the provisions of the Arbitration Act, 1979, which was enacted, inter alia, to maintain that situation.

Attestation
Page for the parties to sign the Contract and for the signatures to be witnessed.

The signatories must have authority to sign for this purpose to provide validity to the contract.

Usually as specified in the Tender document, the Contract may be executed:

(a) Under Seal,

or

(b) Under Hand.

Limitation Act, 1939 (Statute of Limitations) - actions

become 'statute barred' after a specified time lapse from when the cause of action accrued:

(a) Under Seal - 12 years

(b) Under Hand - 6 years

In cases involving personal injury, the period of limitation is 3 years from the accrual of the cause of action, which arises when the damage is known or reasonably ought to have been discovered. The situation is more complicated in instances of developing/continuing illness or incapacity, as may occur with some industrial injuries.

Under contract, normally the cause of action accrues when the contract is broken.

In tort, the accrual of the cause of action has been considered in a variety of cases. The situation following the House of Lords' ruling in Pirelli General Cable Works Limited v. Oscar Faber and Partners (1983) is that the cause of action accrues when relevant and significant damage first occurs, whether it is discoverable or not. An exception may occur where, say, a building is 'doomed from the start' although how and where such a situation applies is unclear; in such a case time begins to run when the building is built.

Exceptions to the principles of time beginning to run occur where fraud is present, such as to conceal damage or a possible cause of future damage. Following Applegate v. Moss (1971) and S32 of the Limitation Act, 1980, where fraud is present, the limitation period commences at the time when the damage either is discovered or reasonably ought to have been discovered.

See also: London Borough of Bromley v. Rush & Tompkins Ltd. (1985).

In Billam v. Cheltenham B.C. (1985), it was held that the commencement of the limitation period may be delayed due to ineffective remedying of defects.

Unless the facts show otherwise, there is no contract until the Articles of Agreement have been executed.

William Lacey (Hounslow) Ltd. v. Davies (1954).

If any work is done prior to this, in expectation of a contract being executed and a contract is not brought into existence, a claim for quantum meruit may result. If, however, a contract is subsequently executed, the provisions for valuation and payment will apply retrospectively to the work executed.

Tollope & Colls Ltd. v. Atomic Power Construction Ltd. (1963).

However, where the terms of the contract have been largely agreed and something like a letter of intent has been recognised as authority for the work to commence, even in the event of a non-execution of a formal contract, a contract will be presumed, the terms being as originally envisaged. This will apply generally (but note responsibilities for work execution, payment, determination, etc.).

Courtney & Fairbairn Ltd. v.Tolaini Bros. (Hotels) Ltd. (1975).

Other relevant cases are:

McCutcheon v. David MacBrayne Ltd. (1964).
Modern Building (Wales) Ltd. v. Limmer & Trinidad Co. Ltd. (1975).
Brightside Kirkpatrick Eng. Services v. Mitchell Construction (1975).

Note: Winn v. Bull (1877) - A document which contains a statement 'subject to contract' will postpone the incidence of liability until a formal document has been drafted and signed, or a contract has been properly entered in an appropriate way.

Sherbrooke v. Dipple (1980) - Parties may get rid of the qualification 'subject to contract' only if both expressly agree that the qualification should be expunged or if such agreement must be implied necessarily.

CONDITIONS: PART 1: GENERAL

Clause 1: Interpretation, definitions, etc.

1.1 All clause references relate to the Conditions of Contract unless specifically stated otherwise.

1.2 The Contract is to be read as a whole. Any term is thus subject to qualification by other, usually related and cross-referenced, terms unless specified as to be read in isolation.

This is a general principle of law.

1.3 Definitions essential to the interpretation of the Contract (most are self-explanatory).

The definitions listed refer to the Contract as a whole; those given in individual clauses are in the context of that clause only (e.g. Variation - Clause 13.1).

Following the ruling by the Court of Appeal in <u>J. Jarvis Ltd. v. Rockdale Housing Association (1986)</u> the meaning of 'the Contractor' in JCT 80 (and similar contracts) includes the servants and agents of the Contractor (i.e. those through whom the organisation operates). It is probable that the definition does not include sub-contractors.

Clause 2: Contractor's obligations

2.1 Primary obligation for the Contractor to execute and complete the Works in accordance with the Contract Documents.

If the quality is to be the subject of the opinion of the Architect, it must be to his reasonable satisfaction (something the B.Q. should really define).

Thus the Architect cannot demand better quality than is stated in the Contract Documents, as amended by A.I.s, regarding quality, etc.

By its nature, this condition excludes any design responsibility on the part of the Contractor - he undertakes to complete the Works in accordance with the supplied design; he does not undertake that the building will be suitable for its intended purpose. Neither does the Contractor undertake that the building will not fail, unless due to standards of workmanship, etc., not in accordance with the Contract. He does not contract to produce a result.

See Cable v. Hutcherson Bros. Pty. (1969) (Australian).

However, the courts now regard Contractors as experts in construction and so any design matter that the Contractor considers dubious should be queried by him, in writing, to the Architect, requesting his instructions and indicating the possible construction problems and/or possible subsequent failure.

Following Duncan v. Blundell (1820), Equitable Debenture Assets Corporation Ltd. v. Wm. Moss and others (1984) and Victoria University of Manchester v. Hugh Wilson & Lewis Wormersly and others (1984), an implied term exists for the Contractor to warn the Architect of design defects known to the Contractor and of such defects the Contractor believes to exist. The belief requires more than mere doubt as to the correctness of the design but less than actual knowledge of errors.

Such action should relieve the Contractor of any liability for subsequent building failure due to inadequate or inappropriate design.

'To complete the Works' refers to practical completion, the point at which, *inter alia*, the defects liability period (D.L.P.) commences. Certain contractual requirements cease at this point also e.g. fire insurance and liquidated damages liability.

The Architect (or Employer) may dictate the working hours, order of work execution and postponement of work *but not the method of executing the work.*

2.2 .1 Contract Bills may not over-ride or modify provisions of the Articles, Conditions or Appendix but may affect them e.g. by noting obligations or restrictions imposed by the Employer.

See M. J. Gleeson Ltd. v. London Borough of Hillingdon (1970).

John Mowlem & Co. Ltd. v. British Insulated Callenders Pension Trust Ltd., and S. Jampel & Ptnrs. (1985): (B.Q. attempted to contain a performance specification for watertight concrete; consulting engineers used for structural design.) There must be a very clear contractual condition to render a contractor liable for a design fault. Design is a matter which a structural engineer is qualified to execute, which the engineer is paid to undertake and over which the Contractor has no control.

.2.1 Contract Bills - to have been prepared in accordance with S.M.M.6.

Any departure from this must be specified in respect of each item or items - usually in the Preambles.

.2 Any error in the B.Q. or any departure from S.M.M.6 not specified - not to vitiate the Contract but to be corrected and the correction treated as a Variation in accordance with Clause 13.2.

Note: If the error or departure were sufficiently great so as to change the entire basis or nature of the Contract it is probably that it *would* vitiate the Contract.

See Pepper v. Burland (1792).

The errors are those concerning B.Q. preparation by the P.Q.S., not pricing errors by the Contractor.

2.3 Discrepancies in or divergencies between documents.

If the Contractor finds a discrepancy between any of the specified documents - drawings, B.Q., A.I.s (except Variations), schedules, levels - Contractor to give the Architect written notice specifying the discrepancy and the Architect shall issue instructions to solve the problem.

'*If*' indicates that the Contractor is not bound to find any discrepancies but specifies what the Contractor must do should he make such a discovery.

Again pricing errors by the Contractor are *not* covered nor is his misinterpretation of the design, unless the design is ambiguous, in which case he may claim.

However, the Contractor is considered to be an expert in construction and so the Courts will require him to execute the appropriate duty of care in such things as cross-checking Contract Documents and drawings.

Thus this situation is by no means crystal-clear; the Contractor should check and only in extreme cases place reliance on this Clause to avoid any liability.

However, in Stanley Hugh Leach Ltd. v. L. B. Merton (1985), the judge accepted that the Contractor was not obliged to check drawings to seek discrepancies or divergencies so as to impose a duty on the Architect to provide further information.

Normally the Contract provisions will prevail over other Contract Documents e.g. B.Q.

See Gold v. Patman & Fotherington (1958).

Supply of Goods & Services Act (1982): implied term that a supplier, acting in the course of a business, will carry out the service with reasonable skill and care.

Clause 3: Contract Sum—additions or deductions—adjustment—Interim Certificates

Where the Contract provides for the Contract Sum to be adjusted, as soon as any such adjustment amount has been determined, even if partial (or even, presumably, 'on account') it may be included in the computation of the following Interim Certificates.

This is a distinct aid to the Contractor's cash flow, the 'life-blood' of the industry, by requiring even estimated, partial or 'on account' Valuations of extras to be included in the following (Valuation and) certification for payment.

Clause 4: Architect's instructions

4.1 .1 Contractor to forthwith comply with all A.I.s which the Architect is expressly empowered by the Conditions to issue.

The only exception is where a Variation is involved to which the Contractor has objected in writing to the Architect.

Valid A.I.s:

2.3	discrepancies, divergencies between documents
6.1.3	statutory requirements at variance with the work
7	Contractor's incorrect setting out
8.3	inspection and testing
8.4	removal of items not in accordance with the Contract
13.2	Variations
13.3	expenditure of P.C. and provisional sums
17.2	defects (schedule)
17.3	making good of defects
23.2	outbreak of hostilities - protective work, etc.
33.1.2	war damage - disposal of debris and damaged work, execution of protection
34.2	antiquities
35.5.2	substitute Clauses 35.11 and 35.12 for NSC/1 and NSC/2, or vice-versa

35.8	failure of Contractor and proposed NS/C to agree
35.10.2 } 35.11.2	nomination of sub-contractor
35.18	failure of NS/C to make good defects - another NS/C to do
35.24.4.1	Contractor to specify default to NS/C and seek further A.I. prior to determination
36.2	nomination of supplier

.2 Receipt by Contractor of written notice from Architect to comply with an A.I.

Non-compliance → Employer engages another to do the work involved in the A.I. and contra charges Contractor accordingly

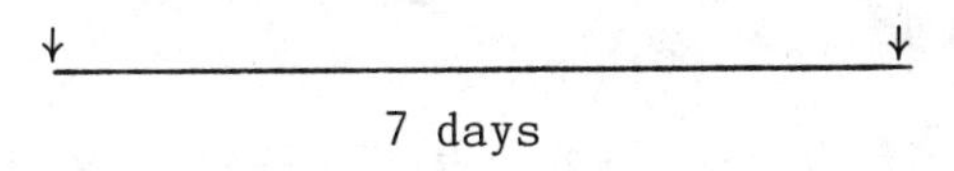

4.2 Contractor receives what purports to be an A.I. from the Architect. Contractor may request Architect to specify the authority (which Clause) for the instruction, in writing.

Architect complies with request.

(In the absence of the instigation of arbitration) Contractor complies with the instruction - deemed, in all cases, to be a valid A.I.

4.3 .1 All instruction issued by the Architect - to be in writing.

.2 Non-written (verbal) instruction - of no immediate effect. Contractor to confirm the instruction to the Architect within 7 days. Architect has 7 days from receipt of confirmation to dissent in writing, if not becomes valid A.I.

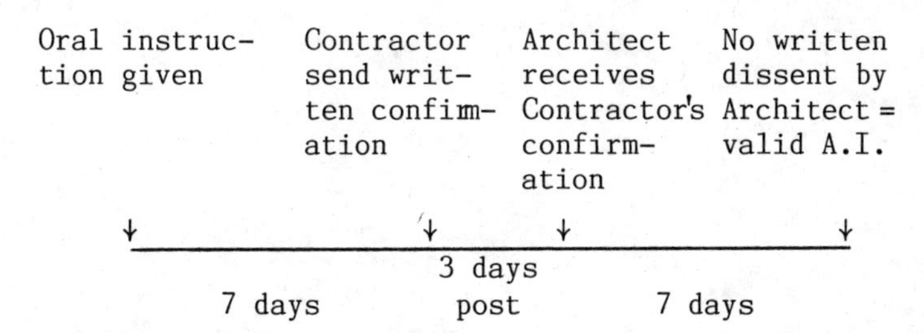

.1 Architect may confirm an oral instruction within 7 days - no need for Contractor to confirm - is valid A.I. from the date of the Architect's written confirmation.

.2 No confirmation of an oral instruction but the Contractor complies - Architect *may* confirm in writing prior to the issue of the Final Certificate - deemed to be a valid A.I. on the date when given orally.

This course of action imposes considerable risk of loss upon the Contractor and much reliance on the Architect's memory!

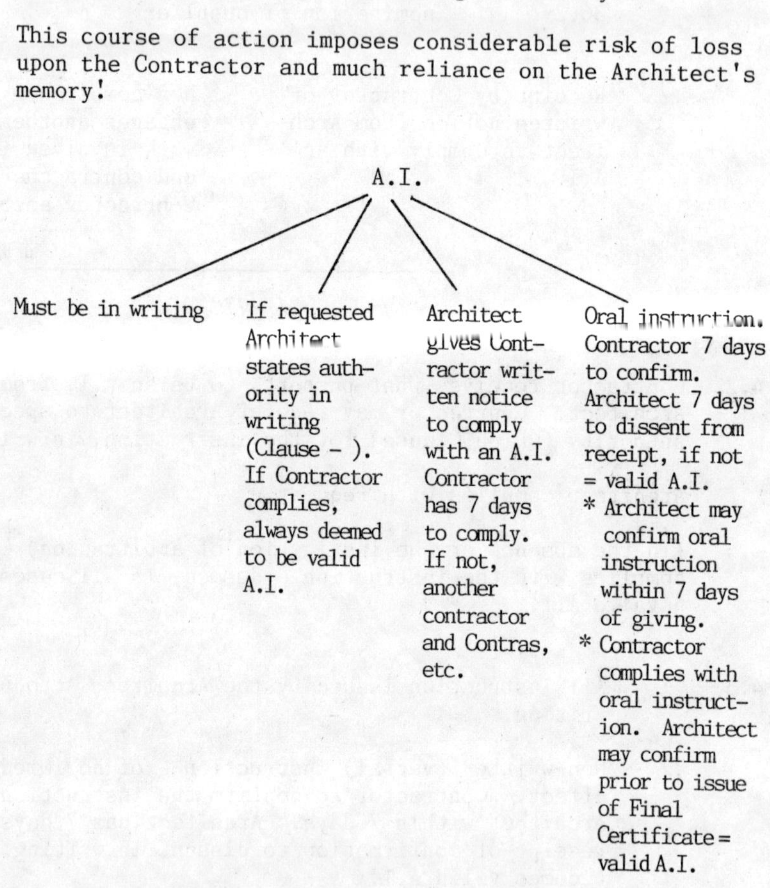

Clause 5: Contract documents—other documents—issue of certificates

5.1 Contract Drawings and Bills to remain in the custody of the Architect or Q.S. and be available at all reasonable times (usually normal working hours) for the inspection of the Contractor and Employer.

5.2 Immediately the Contract is executed, the Architect to provide free to the Contractor:

.1 1 copy of the Contract Documents certified on behalf of the Employer.

.2 2 further copies of the Contract Drawings.

.3 2 copies of the unpriced B.Q.

See also Practice Note 4.

5.3 .1 As soon as possible after the execution of the Contract:

.1 Architect to provide free to Contractor 2 copies of all schedules, etc., necessary for the execution of the Works.

.2 (optional) Contractor to provide free to the Architect 2 copies of the master programme. Amendments made within 14 days of Extension of Time award (Clause 25.3.1) or revised Completion Date due to War Damage (Clause 33.1.3) i.e. to be kept updated in accordance with Architect's Extension of Time award.
(Useful to show when information will be required.)

Note: A network will show the effects of delays far more clearly than a bar chart.

The following points should also be noted:

(a) There is no definitive statement indicating the form of the programme or what it is to show. The type of programme and information shown is thus at the option of the Contractor (in the absence of any agreement denoting specific type of programme and information required). Thus a simple bar chart, a network or any other type of chart will be compliant. Normally it is suggested that the programme should be of a type in common usage, clearly denoting key dates and work sequences, etc. B.Q. may note form of programme required.

(b) If a Contractor disagrees with an extension or other Completion Date modification by the Architect, in order to further his claim it would be perhaps prudent not to amend the programme to comply with the Architext's award but to revise the programme to comply with the Contractor's own estimate of the appropriate Extension of Time and to indicate by this means the procedure for completion by that data.

Indeed, it should be borne in mind that there is no provision in the Contract for the Contractor to amend the programme to take account of the Architect's reduction of an Extension of Time (due to work being omitted).

.2 Nothing in the supplementary documents (schedules, master programme, etc.) may impose any obligation beyond those imposed by the Contract Documents.

5.4 Architect to provide free to the Contractor 2 copies of all necessary drawings and details, supplementary to the Contract Drawings, to enable the Contractor to execute the Works.

See also Practice Note 4.

5.5 Contractor to keep on site and available for the Architect's inspection at all reasonable times:

(a) 1 copy of Contract Drawings.

(b) 1 copy of unpriced B.Q.

(c) 1 copy of Schedules, etc.

(d) 1 copy of master programme.

(e) 1 copy of drawings and details.

5.6 Upon final payment, the Contractor to return to the Architect all drawings, schedules, etc., which bear Architect's name, if requested to do so.

5.7 Limits to use of documents.

Documents to be used for the purposes of the Contract.

Employer, Architect and Q.S. may use the B.Q. rates or prices only for the purposes of the Contract - also may not divulge them except for the purposes of the Contract.

5.8 Normally all Certificates must be issued to the Employer by the Architect who must immediately send a duplicate copy to the Contractor.

(1963 Form - all Certificates issued to the Contractor caused problems, most markedly over payments to NS/C - 14 days from date of receipt of Certificate by the Contractor - payments by Employer - Contractor had to present the Certificate to the Employer to instigate the debt - 14 days to honour Certificate.)

Exception:

35.15.1 failure of NS/C to complete by appropriate date.

Following Stanley Hugh Leach Ltd. v. L. B. Merton (1985) it should be noted that the following terms are implied:

(a) the Employer will not hinder or prevent the Contractor's carrying out his obligations under the Contract or from executing the Works in an orderly and regular way,

(b) the Employerundertakes that the Architect will do all required to enable the Contractor to carry out the work,

(c) where the Architect is required to supply the Contractor with drawings etc. during the course of the work, those drawings etc. must be accurate.

Clause 6: Statutory obligations, notices, fees and charges

6.1 .1 Over-riding obligation to comply with any and all statutory requirements, including the regulations of statutory undertakers.

.2 If the Contractor finds any divergence between the statutory requirements and the Contract Documents, Drawings, etc., or A.I.s he must immediately inform the Architect, in writing, specifying the divergence.

Note: the role of 'if', not 'when', and the implications as previously discussed.

.3 If the Contractor gives notice or by some other means the Architect discovers such divergence, he has 7 days from receipt of any notice in which to issue instructions about the matter. If such instructions require the Works to be varied, this constitutes a valid Variation (Clause 13.2).

.4.1,.2,.3, Emergency compliance with statutory requirements prior to instructions being received:

Contractor to supply necessary materials and to execute necessary emergency work.

Contractor to inform the Architect of the situation and the steps being taken.

Such materials and work are then deemed to be pursuant to an A.I., provided the necessity for the emergency work was due to a divergence as described - valuation will be made accordingly (Clause 13.2).

.5 Provided the Contractor has complied with the requirements regarding divergence, he is not liable where the Works do not comply with any statutory requirements, provided they do comply with the Contract Documents, including Drawings.

<u>Note</u>: This liability avoidance is contractual only, i.e. expressly restricted to liability to the Employer.

No Contractor may construct a building which does not comply with the applicable statutory requirements.

6.2 Fees and charges - Contractor to pay and indemnify the Employer in respect of all statutory fees and charges.

Such fees and charges, including all taxes except V.A.T., are added to the Contract Sum unless they:

.1 arise due to a statutory undertaker being an NSup or NS/C.

.2 are prices in the Contract Bills.

.3 are a provisional sum in the Contract Bills.

6.3 The contractual provisions regarding assignment and sub-letting (Clause 19) and NS/Cs (Clause 35) do not apply to statutory undertakers carrying out their statutory obligations. For this purpose they are not Sub-Contractors within the meaning of the Contract.

Clause 7: Levels and setting out of the Works

The Architect is obliged to provide the Contractor with all requisite levels and dimensions for the execution of the Works.

Expressly to provide drawings to enable the Contractor to set out at ground level.

Unless subject to an A.I. to the contrary, the Contractor is responsible for correcting at his own cost any setting out errors for which he is responsible, i.e. his inaccuracies in setting out.

e.g. Architect could issue an instruction permitting an inaccurately set out building to be completed (presumably, however, with an appropriate price reduction by the Contractor).

Trespass: If setting out causes a trespass on adjoining property, the injured party has an action:

(a) against the Contractor if the trespass is caused by setting out error(s) on his part,

(b) against the Employer (who may be able to recover from the Architect) if the trespass is caused by inaccurate drawings or other information supplied by the Architect.

See Kirkby v. Chessum & Sons Ltd. (1914).

Note: The Contractor may be required to insure against the risks involved:

(a) Clause 20.2.
(b) Clauses 20.2 and 21.2.1.

Clause 8: Materials, goods and workmanship to conform to description, testing and inspection

8.1 Bills to describe all materials, goods and workmanship, thus the B.Q. acts as the specification document.

'So far as procurable' modifies this overall requirement - may be best to give alternatives, if applicable - e.g. paint manufacturers list, in the Preambles. If this is not done, the phrase may be deemed to mean the nearest substitute to the specified item, if unavailable.

In such circumstances, it would be prudent to seek the Architect's approval and instructions.

This Clause has led to the wide use of the term 'or equal and approved' - this means the Architect *may* approve other goods, not *must*.

See Leedsford Ltd. v. Bradford Corpn. (1956).

8.2 Architect may require the Contractor to provide vouchers to ensure/prove that the goods and materials comply with the specified requirements as Clause 8.1.

8.3 A.I. may be issued for the Contractor to open up work for inspection or to test goods and materials.

Costs to be added to the Contract Sum or may be provided in the B.Q. (Provisional Sum) - including costs of making good.

However, if the tests etc., prove the work/materials to be defective, the Contractor must bear the costs.

8.4 Architect may issue an A.I. for removal from the site of work, goods or materials not in accordance with the Contract requirements.

Note: (a) The Architect should give the reason for the A.I., as removal is a Variation under Clause 13.1.1.3 where the goods, etc., *are* in accordance with the Contract.

(b) By Clause 27.1.3, the employment of the Contractor may be determined if the defective items are not removed, provided the items are material to the Works.

(c) Defects becoming apparent due to, *inter alia*, Clause 17.2 unspecified materials, etc., must be made good at the Contractor's expense.

Note: Following Holland Hannen and Cubitts (Northern) Ltd. v. Welsh Health Technical Services Organisation (1981), an Architect must take care over instructions issued regarding apparently defective work. The authority given by the Contract, Clause 8.4, is for the Architect to instruct the Contractor to remove the defective work etc.; the Contractor remains under the obligation to complete work in accordance with the Contract.

8.5 Architect may (not unreasonably or vexatiously) issue an A.I. for the dismissal *from the Works* of any person employed thereon. This means removal from that particular project, *not* dismissal from employment.

A person is defined in this context as 'an individual or firm (including bodies corporate)' so firms and individuals may be dismissed.

Note: (a) Interim payments include only items of work properly executed in accordance with the Contract requirements - as this decision of compliance is that of the Architect, in theory the Q.S. may not deduct from his recommendation for payment unless an A.I. as to unsuitability exists.

In such circumstances, however, the Q.S. may be

held not to be fulfilling his role as a professional (expert) and so should include a note of his valuation of any 'suspected defective work'. It is then for the Architect to deduct any appropriate amounts at certification.

See Sutcliffe v. Thackrah (1974).

(b) From several House of Lords decisions, the Contractor gives warranties or implied terms:

(i) that the workmanship will be of a good standard,

(ii) that materials and goods supplied will be free from latent and patent defects (including any under a nomination), and that they

(iii) will be suitable for the purposes for which they are supplied.

These may be expressly excluded, or excluded by the circumstances of the Contract. The Standard Form does *not* refer to these matters and so unless other Contract Documents give specific reference, they will be implied terms.

See Young & Marten Ltd. v. McManus Childs Ltd. (1968) and Gloucestershire C.C. v. Richardson (1968).

Unfair Contract Terms Act, 1977 - in reference to exemption clauses - Contractor should ensure that the above warranties are passed on to suppliers - note that if the Architect insists on nominating a supplier who will not accept these warranties, the Contractor has no liability in respect of the items in question except for patent defects (defects obvious on reasonable examination).

Limits application of exemption clauses in regard to consumer transactions/sales.

Clause 9: Royalties and patent rights

9.1 All sums payable in respect of patented items are deemed to have been included in the Contract Sum by the Contractor where described by or referred to in the Contract Bills.

Contractor to indemnify the Employer against any claim for infringement of patents by the Contractor.

9.2 Where Contractor uses patented items due to his compliance with an A.I., he is not liable for any patent infringements.

Any royalties, damages, etc., so arising to be paid by the Contractor must be added to the Contract Sum.

Note: (a) Where patents are infringed by the Contractor and the articles in question are not referred to in the Contract Bills, the Employer may incur liability, depending on the circumstances, provided the articles were for proper inclusion in the Works.

(b) Prima facie, the Employer is not liable in respect of A.I. provision unless the A.I. constitutes a Variation.

(c) For the implementation of this Clause, nominated items are presumed to be the responsibility of the Architect - the Contractor will not usually be fully conversant with the contents of the nomination at the time of tender - note, however, S.M.M.6 and 1980 Form provisions.

Clause 10: Person-in-charge

Contractor to keep a *competent* person-in-charge on the site (usually taken as normal working hours or actual site working hours). This person is given the power of Agency on behalf of the Contractor in respect of receipt of A.I.s and C.o.W directions - hence the common title 'Site Agent'.

All relevant persons should be informed who the Site Agent is and who is the deputy (the person who will assume the responsibilities in the absence of the Agent). Any changes in these personnel should be notified immediately.

Clause 11: Access for Architect to the Works

Access is to be for the Architect and his representatives and at all reasonable times. This applies to the site and all workshops, etc., including those of Sub-Contractors, where items for the project are being made, so far as procurable by the Contractor.

This will, of course, include the C.o.W., engineers and Q.S. (especially for valuation purposes).

Clause 12: Clerk of Works

Employer is entitled to appoint a C.o.W.

C.o.W. acts solely as an inspector on behalf of the Employer but under the directions of the Architect.

Contractor to give adequate facilities for the C.o.W. to perform his duties.

C.o.W. directions:

(a) given to the Contractor,

(b) no immediate effect,

(c) to be valid must be in respect of a matter about which the Conditions expressly empower the Architect to issue instructions,

(d) confirmed in writing by the Architect within 2 working days of being given to be of any effect - if so confirmed, then from date of confirmation - deemed to be an A.I.

Contractor has no contractual power to object to a C.o.W.

Provided the Architect has properly briefed the C.o.W., the Employer is responsible for him.

Leicester Board of Guardians v. Trollope (1911) shows that the Architect is responsible for his design being followed - not avoided by a C.o.W. being on site.

Clause 13: Variations and provisional sums

See also Practice Note 14.

13.1 Defines Variations:

.1 Alteration or modification of design, quality or quantity of the Works as described in the Contract Drawings and Bills, including:

- .1 addition, omission or substitution of any work
- .2 alteration of any kind or standard of any materials or goods to be used in the Works
- .3 removal from site of materials/goods for the Works which *are* in accordance with the Contract (brought on site by the Contractor).

.2 Alteration, addition or omission of obligations/ restrictions imposed by the Employer in the Contract Bills regarding:

- .1 access to the site or use of any specific parts of the site
- .2 limitations of working space
- .3 limitations of working hours
- .4 the execution or completion of the work in any specific order (*not* allowed under 1963 Edition).

Note: If sections are to be completed by specific dates in advance of the overall completion date, then the Sectional Completion Supplement must be used.

.3 Specifically excludes omitting Contractor's measured work and the substitution of a Nominated Sub-Contractor to execute that work.

Note: (Including 1963 Edition) Must be a P.C. or provisional sum for a valid nomination.

Any omissions must be genuine, otherwise the Contractor may claim loss of profit for the 'omissions'.

13.2 Architect may issue A.I.s requiring Variations. Architect *may* sanction in writing any Variation made by the Contractor *not* pursuant to an A.I.

'No Variation required by the Architect or subsequently sanctioned by him shall vitiate this contract.'

This does *not* mean that the Architect may order any changes he likes and still maintain the Contract intact.

An *excessive* Variation ordered by an A.I. would not be regarded as a Variation under the Contract, thus payment would be on *quantum meruit*.

Lord Kenyon in Pepper v. Burland (1792):

'If a man contracts to do work by a certain plan, and that plan is so entirely abandoned that it is impossible to trace the Contract, and to what part of it the work shall be applied, in such case the workman shall be permitted to charge for the whole work done by measure and value, as if no Contract had ever been made.'

13.3 Architect to issue A.I.s regarding:

.1 expenditure of provisional sums included in the Contract Bills,

.2 expenditure of provisional sums included in a Sub-Contract (applied where nomination occurs from expenditure of a provisional sum as Clause 13.3.1).

13.4 .1(a) Valuation of Variations - by the Q.S. - in accordance with Clause 13.5 (unless otherwise agreed by the Contractor and the Employer - this would constitute a Variation of the Contract and so the parties to the Contract must agree to it).

Note: The payment rules are applicable only to Variations as defined by the Contract. If in doubt, therefore, the Contractor should implement the authority for an A.I. procedure as Clause 4.2.

See Myers v. Sarl (1860): Drawings, even if prepared in the Architect's office and stamped, must be signed by the Architect to be valid authority to execute work. Possibly, however, the signature of a clerk preparing the drawings on behalf of the Architect would suffice.

(b) Valuation of Variations - Nominated S/C's work - to be in accordance with the relevant provision of NSC/4 or 4a (Sub Contract forms); unless otherwise agreed by NS/C and Contractor, with Architect's approval.

.2 Valuation of work constituting a P.C. Sum arising due to A.I.s regarding expenditure of a provisional sum for which the Contractor has had a tender accepted (Clause 35.2) must be valued in accordance with that successful tender, *not* the normal Variation provisions.

13.5 .1 Variations constituting additional or substituted work which can be valued by measurement - valuation to be of:

.1 *similar* character, executed under *similar* conditions, does not *significantly* change quantity of work in Contract Bills - Rates and Prices in B.Q. to be used.

.2 similar character, *not* similar conditions and/or significantly changes quantity of work in B.Q. - Rates and Prices in B.Q. to form basis for valuation (i.e. *pro rata* prices).

.3 *not* of similar character to B.Q. - fair rates and prices.

Rather contentious - can varied work really be of similar character? What is a significant change in quantity? How much to *pro rata*? What is a fair valuation? Perhaps a guide is to read 'identical' for 'similar'.

.2 Variations requiring omission of work from B.Q.: valuation of work omitted to be at B.Q. rates.

.3.1 Measurement to be in accordance with the principles used to prepare the B.Q. (S.M.M.6).

.2 Allowances to be included in respect of any appropriate percentages or lump sums in B.Q. (e.g. profit, attendances on NS/Cs).

.3 Appropriate adjustment to be made for preliminaries.

.4 Where additions or substitutions cannot properly be valued by measurement, valuation shall comprise:

.1 Prime cost of the work plus percentage additions as set out by the Contractor in the B.Q. (usually separate percentages for labour, plant, materials). Prime cost - calculate in accordance with 'Definition of the Prime Cost of Daywork carried out under a Building Contract' current at the Date of Tender.

.2 Specialist trades definition of Prime Cost of Daywork to be used as appropriate.

(Electrical Contractors' Association
Electrical Contractors' Association of Scotland
Heating and Ventilating Contractors' Association).

Vouchers stating:

time daily spent on the work,
workmen's names (usually including grade),
plant,
materials,

must be sent to Architect, or his authorised representative for verification not later than the end of the week following that during which the work is executed. Verification is usually by signature of the appropriate person - if in doubt send to the Architect.

.5 If a Variation *substantially* changes the conditions under which other work is executed, that other work may be valued in accordance with the rules for valuing Variations.

.6 Excluding additions, omissions and substitutions of work, the valuation of work or liabilities directly associated with a Variation which cannot be reasonably valued as above, shall be the subject of a fair valuation, e.g. part load delivery charges where the bulk of the items in question have already been delivered.

13.6 An allowance for disturbance to the regular progress of the

Works and/or for any direct loss and/or expense may be included in the valuation of Variations under Clause 13.5 only where it is *not* reimbursable under any other Contract Clause.

Such 'disturbance cost' will usually be considered under Clause 26 - the provision of Clause 13.6 being used only as a 'last resort'.

13.7 Q.S. to allow Contractor to be present and take all necessary notes when Q.S. is measuring work to value Variations.

13.8 Contract Sum to be adjusted to take account of all valuations of Variations.

Thus, where a Variation has been executed but not measured and valued, it is reasonable to include an 'on account' valuation in interim payments until formal valuation has been finalised.

Note: The valuation of Variations, howsoever arising, is to be executed by the Q.S. (including provisional sum items) unless the Employer and Contractor themselves agree something different. The Architect is involved only where a Contractor claims for reimbursement of loss/expense due to the regular progress of the Works being materially affected (disturbed).

Clause 14: Contract Sum

14.1 'The quality and quantity of the work included in the Contract Sum shall be deemed to be that which is set out in the Contract Bills'.

Note: Clause 2.2 - The Bills of Quantities must be in accordance with S.M.M.6. Any items not so included must be specifically noted with an indication of how measured, usually indicated thus:

'Notwithstanding S.M.M. Clause No. , work item, is measured description of how measured for B.Q.'

Such a statement will usually occur in the Preambles section of the B.Q.

Clause 14.1 is upheld in its expressed limitation of authority of the B.Q. by:

English Industrial Estates Corporation v. George Wimpey & Co. Ltd. (1973)

M. J. Gleeson Ltd. v. L. B. of Hillingdon (1970).

14.2 Contract Sum fixed as a lump sum; the only adjustments permissible are those set out in the Conditions of Contract (e.g. valuation of Variations).

Subject to Q.S.'s preparation errors which are covered by

Clause 2.2.2.2, any errors (arithmetic, etc.) are deemed to have been accepted by the parties and are non-adjustable.

Note: (a) Provisions of Code of Procedure for Single Stage Selective Tendering, 1977.

(b) Professional responsibilities, especially negligence.

(c) J.C.T. Standard Form is a Lump Sum Contract with provision for interim payments.

Clause 15: Value added tax—supplemental provisions

15.1 Defines V.A.T. - introduced by Finance Act 1972.
Under the control of the Customs and Excise.

15.2 Contract Sum is always *exclusive* of V.A.T.

15.3 Where any items become *exempt* from V.A.T. after the Date of Tender - Employer to pay the Contractor the loss of input tax he would otherwise have recovered.

J.C.T. Practice Note 6 is a useful guide to the V.A.T. provisions.

Clause 16: Materials and goods unfixed or off-site

16.1 Unfixed materials/goods delivered to, placed on or adjacent to the Works and intended therefore may be removed only for use on the Works, unless removal has been agreed in writing by the Architect. Such consent not to be unreasonably withheld.

Unfixed materials/goods (as above) - value of which has been excluded in an Interim Certificate which has been paid are the property of the Employer. The Contractor is in the position of a bailee, expressly responsible for loss or damage to them; but subject to Clause 22B or C if applicable (fire, etc., Employer to insure).

Until an Interim Certificate, as denoted above, is honoured, the property in the goods remains with the appropriate party as per the sale agreement, governed by the Sale of Goods Acts (usually the Contractor).

Following Dawber Williamson Roofing Ltd. v. Humberside County Council (1979); under a sub-contract in which no express terms cover the passing of property in materials/goods on site, the title does not pass until the materials/goods are fixed.

See Stansbie v. Troman (1948): where items on site are owned by the Employer, common law requires the Contractor to take reasonable care to protect them from damage, theft, etc. This probably also applies to Sub-Contractors' items on site.

16.2 Applies to materials/goods stored off site - often applies to nominated items due to their high value. Once the Employer has paid the Contractor for them (Clause 30.3) and so the property has passed to the Employer, the Contractor has responsibilities:

(a) to remove them only for incorporation in the Works

(b) for costs of storage, handling, insurance until delivered to the Works, whereupon Clause 16.1 applies

(c) for loss or damage.

Note: S.16 Sale of Goods Act - property in unascertained goods cannot pass to the purchaser unless and until the goods are ascertained.

Naturally many of these responsibilities are passed on by the Contractor to the supplier or Sub-Contractor in most instances.

In the absence of any express terms in a building Contract, materials on site do not become the property of the Employer until they have been made part of the project; 'fully and finally incorporated into the Works' is a common phrase in this regard.

Clearly, in the 1980 J.C.T. standard form, this general principle is over-ridden by the express provisions of the Contract - Clause 16.

See also Practice Note 5.

Clause 17: Practical completion and Defects Liability

17.1 As soon as the Architect considers that Practical Completion of the Works has been achieved, he must issue a Certificate to that effect. For all the purposes of the Contract, Practical Completion is deemed to have occurred on the day named in that Certificate.

Practical Completion is not defined in the Contract but is normally understood to be when the Works are complete for all practical purposes, any outstanding items of work being of only a minor or remedial nature such that they would not materially affect the proper functioning of the building.

Note: Practical Completion is distinct from and not applicable to the doctrine of Substantial Completion where, if a party can show that he has 'substantially performed' his obligations, he can successfully sue for the price under an entire or lump sum contract provided he gives appropriate credit for any deficiences in the whole, including uncompleted obligations.

See Appleby v. Myers (1867) - following Cutter v. Powell (1975) - Blackburn J:

'There is nothing to render either it illegal or absurd in the workman to agree to complete the whole, and to be paid when the whole is complete, and not till then.'

If a builder failed to complete an entire contract he could

claim neither *quantum meruit* nor in equity.

Sumpter v. Hedges (1898) - A. L. Smith L. J.:

'The law is that where there is a contract to do work for a lump sum, until the work is completed the price of it cannot be recovered.'

H. Dakin & Co. Ltd. v. Lee (1916): Under a lump sum building contract, defects or omissions amounting only to a negligent performance (not abandonment of the contract, etc.) did not preclude a successful claim for the contract sum less only the amount necessary to make the work accord with the specification.

Naturally the question is one of degree and for this rule to apply the omissions or defects should be of a relatively minor nature.

Hoenig v. Isaacs (1952) - Denning L. J.:

'It was a lump sum contract, but that does not mean that entire performance was a condition precedent to payment. Where a contract provides for a specific sum to be paid on completion of specified work, the courts lean against a construction of the contract which would deprive the contractor of any payment at all simply because there are some defects or omissions.'

Thus, it may be concluded that the J.C.T. Standard Form 1980 Edition is a lump sum contract with provision for Interim Payments.

Note: following observations in Dakin v. Lee and Hoenig v. Isaacs:

(a) the builder cannot recover if he abandons the contract (subject to the express terms of the agreement)

(b) contracts which provide for retention money to be paid on completion might require entire performance in the strict sense - but probably in relation only to the retention releases rather than Interim Payments under the Contract.

17.2 The Appendix provides for the usual 6 months D.L.P. to be varied.

Defects, shrinkages and other faults which appear within the D.L.P. and are due to materials and workmanship not in accordance with the Contract must be:

(a) specified on a schedule of defects by the Architect, to be delivered as an A.I. to the Contractor not later than 14 days from the expiration of D.L.P.

(b) made good by the Contractor at his own cost (unless subject to an A.I. to the contrary) and within a reasonable time.

17.3 Despite the provision of Clause 17.2, such defects, etc., including damage caused by frost prior to Practical Completion, may be the subjects of an A.I. for their making good - Contractor to comply within a reasonable time and normally at his own cost.

No such A.I.s may be issued after delivery of the schedule of defects or 14 days from expiration of D.L.P.

This Clause enables individual defects, usually of a more major nature, to be required to be made good prior to the issue of the full defects list. There is no provision for an interim defects list except by this A.I. provision but the issue of such a list is quite common in practice.

Thus it is sensible and usual for the defects schedule to be prepared and issued as late as possible such that all defects arising may be properly included.

Any defects which appear after the expiration of D.L.P. are not covered by the Contract and so any action for their making good would have to be at common law.

Note: Statute of Limitations provisions.

17.4 When the defects specified by the A.I.s and/or the schedule of defects have, in the Architect's opinion, been made good, he must issue a Certificate of Completion of Making Good Defects.

This Certificate is a pre-requisite for:

(a) Final Certificate - Clause 30.8,

(b) release of the balance of retention - Clause 30.4.1.3

17.5 Contractor must make good only frost damage which is due to frost prior to Practical Completion. If frost damage becomes apparent after Practical Completion the Architect must certify that such damage is due to frost which occurred prior to Practical Completion for the Contractor to be required to make it good under the Contract. If no such Certificate is issued, the Contractor is entitled to claim payment for the work involved.

Clause 18: Partial possession by Employer

18.1 Prior to the date(s) of issue of the Certificate(s) of Practical Completion, the Employer may take possession of any part(s) of the Works provided the consent (not unreasonably withheld) of the Contractor has been obtained. Upon such taking of possession, the Architect must issue a written statement to the Contractor, on the Employer's behalf, which identifies the part(s) of the Works taken into possession ('relevant part') and which specifies the date on which the Employer took possession ('relevant date').

In such instances, the following must occur:

.1 Practical Completion is deemed to have occurred and the D.L.P. to have commenced on the relevant date for the relevant part, for the following purposes only:

(a) Clause 17.2 - schedule of defects

(b) Clause 17.3 - A.I.s re defects

(c) Clause 17.5 - damage by frost

(d) Clause 30.4.1.2 - Retention (½ release).

.2 Certificate of Completion of Making Good Defects of the relevant part to be issued at the appropriate time (see usual provisions regarding the issue of this Certificate).

.3 The obligation to insure the relevant part under Clause 22A (Contractor insuring) or Clause 22B.1 or 22C.2 (Employer insuring) terminates from the relevant date. Where Clause 22C applies, the obligation of the Employer to insure under Clause 22C.1 includes the relevant part from the relevant date.

.4 Liability to pay liquidated damages in respect of the relevant part ceases on the relevant date. Any such liability (Clause 24) in respect of the remainder of the Works which arises on or after the relevant date is reduced on a pro-rata basis.

(The value of the relevant part pro-rata the Contract Sum.)

English Industrial Estates Corpn. v. George Wimpey & Co. Ltd. (1972) stresses the importance of the Architect's Certificate.

See also Practice Note 22.

Clause 19: Assignment and Sub-Contracts

See also Practice Note 9.

Note: the difference between assignment and sub-letting.

Assignment - the transfer of one party's total rights and obligations under a contract to a third party. The consent of the other party to the original contract is often required (e.g. debt factoring). Obligations cannot be transferred unless attached to some right(s).

Sub-letting - where one party's contractual obligations are carried out on his behalf by a third party. The original parties to the contract retain full rights and responsibilities under that contract. The third party will often be in a contractual (Sub-Contract) relationship with the party for whom he carried out obligations vicariously.

19.1 Neither party may assign the Contract without the written consent of the other.

19.2 The Contractor must have the written consent of the Architect to sub-let any part of the Works. Such consent must not be unreasonably withheld.

Defines a Domestic Sub-Contractor (one who is not Nominated).

Note: L.A. common requirements to sub-let only to local firms - usually within a prescribed area.

There is no requirement for the Architect to approve Domestic Sub-Contractors.

19.3 .1 Work measured in the Contract Bills and priced by the Contractor, where the B.Q. requires that work to be carried out by a person selected by the Contractor from a list contained in the B.Q.

.2.1 The list must contain at least 3 persons. The Employer (or the Architect on his behalf) or Contractor may, with the consent of the other party, add persons to any such list prior to the execution of a binding Sub-Contract in respect of that work.

.2 If, prior to the execution of the Sub-Contract, less than 3 persons named in a list are able and willing to carry out the work:

either - The Employer and Contractor by agreement add further names to the list to make it at least 3 names long,

or - the work may be carried out by the Contractor who may sub-let it to a Domestic Sub-Contractor.

.3 A person selected from such a list is to be a Domestic Sub-Contractor.

Note: If a list is altered by having names added, such additions must be inserted in the B.Q. and initialled by the Employer and Contractor as they represent changes in the terms of the Contract.

19.4 .1 Any domestic sub-contract must provide for the employment of that Domestic Sub-Contractor to be determined immediately that the employment of the main Contractor is determined under the Contract.

.2 Any domestic sub-contract must provide that:

.1 Unfixed materials/goods on site of the Domestic Sub-Contractor must not be removed except for use on the Works, unless the Contractor has consented in writing to such other removal. The consent must not be withheld unreasonably.

.2 Where the Employer has paid the Contractor in accordance

with Clause 30 for unfixed materials/goods on site, those materials/goods are deemed to be the property of the Employer. The Domestic Sub-Contractor may not deny that the property in the materials/goods has passed to the Employer.

.3 If the Contractor pays the Domestic Sub-Contractor for unfixed materials/goods on site prior to the Employer's properly paying the Contractor for them, the property in those materials/goods, passes to the Contractor upon the payments being made to the Domestic Sub-Contractor.

.4 The Operation of Clauses 19.4.2.1 to 3 is without prejudice to the provisions of Clause 30.3.5.

Clause 19.4.2 is intended to overcome difficulties found to exist over the title to unfixed materials/goods on site in Dawber-Williamson Roofing Ltd. v. Humberside County Council (1979), or in instances where domestic sub-contracts include retention of title provisions ('Romalpa Clauses' - Aluminium Industrie Vaassen v. Romalpa (1976)). Clauses 19.4.2.2 and .3 seek to ensure that the Employer will have good title to unfixed materials/goods on site for which payment has been made under the provisions of the main Contract, even in cases where the Contractor does not have good title to those materials/goods to pass on - this is the objective of the 'not deny' stipulation in Clause 19.4.2.2.

19.5 .1 Provisions for Nominated Sub-Contractor - see Part II of the Contract.

.2 Unless the Contractor is acting as a Nominated Sub-Contractor (Clause 35.2) he is not required to do any work which is to be carried out by a Nominated Sub-Contractor.

Bickerton v. N. W. Regional Hospital Board (1969): If a first nomination fails before the work under a P.C. Sum is completed there is a duty upon the Architect to re-nominate.

Note: Non-compliance with the sub-letting provisions (Clause 19.2 etc.) is a ground for determination of the Contractor's employment (Clause 27.1.4.).

British Crane Hire Ltd. v. Ipswich Plant Hire Ltd. (1975): If there is a 'course of dealing' between the parties (frequent transactions probably using a standard contract) or if the terms of agreements are standard (e.g. national plant hire agreement), even if not specifically included in forming a contract, they may be implied to give the relationship 'business efficacy'.

Clause 19A: Fair Wages (L.A. Form only)

Result of the House of Commons Fair Wages Resolution of 14 October 1946. This resolution was withdrawn by the House of Commons on 21 September 1983.

The result of the withdrawal is, inter alia, to render Clause 19A.3 (reference re question about observation of Clause 19A provisions to an independent tribunal for decision via the Minister of Labour) inapplicable and so Clause 19A.3, if included in the printed contract form should be deleted.

The overall requirements are for the Contractor to pay wages in keeping with the level of wages for the various trades applicable in the area and to apply similar constraints in respect of working hours, conditions, etc.

The Contractor warrants that, as far as he knows, he has complied with the requirements of this Clause for at least 3 months prior to the date of his tender for the Contract.

The Contractor must allow his employees to be members of trade unions.

The Contractor must display a copy of Clause 19A during the Contract in every appropriate workplace together with a copy of any appropriate wage agreements (local or national). The latter, if not displayed, must be available for inspection.

The Contractor is responsible for observance of this Clause by all his Sub-Contractors.

The Contractor to keep proper wage and time records which are subject to inspection by the Employer (or his representative for that purpose).

The Employer or Architect may require proof of compliance.

Clause 20: Injury to persons and property and indemnity to Employer

See Also Practice Note 22.

20.1 The Contractor to be liable for and must indemnify the Employer against any expense, liability, loss, claim or proceedings at statute or common-law for personal injury or death due to the carrying out of the Works.

The only exception is the extent to which negligence on the part of:

(a) the Employer

(b) any person for whom the Employer is responsible

(c) any person(s) employed or engaged by the Employer to whom Clause 29 refers

caused the injury/death. (Thus if the Employer's negligence is 30% to blame for the injury, the indemnity covers 70% of the Employer's loss).

20.2 Similar to Clause 20.1, but:

(a) refers to real or personal property

(b) excludes the Works etc. as Clause 20.3

(c) excludes damage to existing structures, contents etc. under Clause 22C.1 - Employer to insure

(d) damage must be due to any negligence, breach of statutory duty, omission or default of the Contractor or persons for whom the Contractor is responsible. Includes all persons authorised to be on the site except:

(i) the Employer
(ii) persons employed, engaged or authorised by the Employer
(iii) persons employed, engaged or authorised by any local authority
(iv) persons as (iii) but employed etc. by a statutory undertaker and who are executing work solely in pursuance of its statutory rights or obligations.

20.3.1 'Property real or personal' in Clause 20.2 subject to partial possession, excludes the Works, materials on site and work executed up to and including the earlier of:

(a) the date of issue of the Certificate of Practical Completion, or

(b) the date of determination of the employment of the Contractor (whether or not disputed) under Clauses 27 or 28 or 22C.4.3, if applicable.

Thus the Contractor is responsible for these items and so should effect appropriate insurance.

.2 If partial possession has occurred under Clause 18, the relevant part is excluded from the Works or work executed under Clause 20.3.1.

The purpose of indemnity is to protect against legal responsibility or to compensate; insurance proves a fund to enable the indemnifying party to make any payments which may arise. Insurance does not affect the obligations of the parties. Thus, it is usual for indemnity and insurance provisions to be considered together, the former apportioning risks and the latter dealing with the settlement of claims in respect of prescribed liabilities.

Law Reform (Married Women and Tortfeasors) Act, 1935:

Clause 20 covers tortious liability, thus the Act is of relevance. If there were joint (two or more) tortfeasors and only one were sued by a plaintiff and thereby had to pay damages, that tortfeasor could recover from the other, joint tortfeasors a contribution to the damages paid in proportion to the liabilities of each tortfeasor in respect of the tortious act.

Such does not apply if one joint tortfeasor is to indemnify the other(s); the party who is to indemnify the other(s) bears the full liability.

Here the Contractor must indemnify the Employer.

See also:

A. E. Farr Ltd. v. Admiralty (1953)

A. M. F. International v. Magnet Bowling Ltd. & G. P. Trentham Ltd. (1968).

Clause 21: Insurance against injury to persons or property

See also Practice Note 22.

21.1.1.1 The Contractor must take out and maintain and require Sub-Contractors to take out and maintain insurance in respect of liabilities placed upon them under Clauses 20.1 and 20.2, i.e.:

(a) for personal injuries etc., arising from the Works (except if due to Employer's etc. negligence)

(b) for damage to real or personal property arising from the Works due to the negligence etc. of the Contractor etc. Note exclusions of persons causing the damage under Clause 20.2; item (d).

.2 Insurance for injuries to Contractor's and Sub-Contractor's employees must comply with the provisions of the Employers Liability (Compulsory Insurance) Act 1969 including any amendments etc. to that act.

The minimum insurance cover required for all other claims under Clause 21.1.1.1 for any one occurrence or series of occurrences arising out of one event is stated in the Appendix.

Beyond the scope of the provisions of the Contract, the Contractor is required to comply with all statutory provisions, including those regarding insurance, to be effected by an employer.

21.1.2 The Contractor and Sub-Contractors must send documentary evidence to the Architect for inspection by the Employer that the insurances required by Clause 21.1.1.1 have been taken out and maintained whenever reasonably required to do so by the Employer.

On any occasion, the Employer may require (not unreasonably or vexatiously) such documentary evidence to be relevant policy (policies) and premium receipt(s).

.3 If the Contractor defaults in:

(a) taking out the insurance, or
(b) maintaining the insurance, or
(c) causing any Sub-Contractor to take out and maintain the insurance

as required under Clause 21.1.1.1, the Employer may effect the appropriate insurance and recover the premium amounts (paid or payable) from the Contractor:

(a) by deduction from payments to the Contractor under this Contract, or
(b) as a debt of the Contractor.

21.2.1 If the Appendix states that insurance under this Clause (21.2.1) may be required by the Employer, upon being so instructed by the Architect, the Contractor must take out and maintain a Joint Names Policy (the joint names being those of the Employer and the Contractor) for the amount of indemnity stated in the Appendix. The insurance is against:

'... any expense, liability, loss, claim or proceedings which the Employer may incur or sustain by reason of any damage to any property other than the Works and Site Materials caused by collapse, heave, vibration, weakening or removal of support or lowering of groundwater ...'

due to the execution of the Works.

Exceptions are injury or damage:

.1 for which the Employer is liable under Clause 20.2

.2 due to errors or omissions in the design

.3 which reasonably can be foreseen to be inevitable (Rylands v. Fletcher (1868))

.4 for which the Employer should insure under Clause 22C.1, if applicable

.5 arising from war risks or the Excepted Risks (see definition of Excepted Risks in Clause 1.3).

This insurance covers the Employer's potential liability as a joint tortfeasor - if the Contractor wishes to insure in respect of his potential liabilities in this regard, he must do so in addition to the insurance required by this Clause.

Under a joint names policy, the insurers cannot recover payment to one insured party from the other insured party where the latter's action caused the loss.

See Gold v. Patman & Fotheringham Ltd. (1958).

The insurance also will provide protection for the Employer where there is no negligence on the part of the Contractor etc.

It would be prudent for the insurance under Clause 21.2.1 and that effected by the Contractor in respect of Clause 20.2 to be with the same insurer, thereby avoiding potential argument between insurers about which is liable.

.2 The Clause 21.2.1 insurances to be placed with insurers approved by the Employer. Contractor to send policy (policies) and premium receipt(s) to the Architect for deposit with the Employer.

.3 Amounts paid by the Contractor to effect these insurances are to be added to the Contract Sum.

.4 If the Contractor defaults in respect of these insurance requirements, the Employer may effect the insurance required.

21.3 Damage and injury caused by Excepted Risks are excluded from:

(a) indemnity provisions of Clauses 20.1 and 20.2

(b) insurance requirements of Clause 20.1.1

Thus, the Excepted Risks are assumed by the Employer.

Clause 22: Insurance of the Works

See also Practice Note 22.

If it is not possible to effect insurance in respect of all the 'Specified Perils' (see Clause 1.3 for definition), the matter should be resolved at Tender stage and the Contract amended accordingly (this should not be necessary).

Which of the three alternative Clauses will be used depends upon the circumstances:

22A - Contractor to insure) new building
22B - Employer to insure)
22C - Employer to insure - alterations/extensions to an existing structure

Each insurance comprises a Joint Names Policy for All Risks Insurance for the Works. Clause 22C insurance further comprises a Joint Names Policy for insurance of the existing structures and their contents owned by the Employer or which are the Employer's responsibility against loss or damage caused by the Specified Perils.

22.1 Which Clause, 22A, 22B or 22C is to apply must be stated in the Appendix.

Note: A new building, if attached to an existing building by a link bridge or similar connection, depending on the nature of the connection of the buildings, may be deemed to be an extension to the existing building for insurance purposes.

22.2 Sets out relevant definitions of terms, namely:

All Risks Insurance - which gives cover against any physical loss or damage to work executed and Site Materials.

Costs of repairing, replacement or rectification are excluded in respect of:

1 defects due to
 .1 wear and tear
 .2 obsolesence
 .3 deterioration, rust or mildew;

2 work executed or Site Materials lost or damaged through defective design or work execution etc. of those items, and any work executed which is lost or damaged as a consequence of such failure where the lost or damaged work relied for support or stability on the items which failed;

Note: Site Materials do not feature in the second part of this sub-Clause.

3 loss or damage caused by or arising from
 .1 any consequence of war, invasion etc., nationalisation, requisition, loss, destruction or damage to property by or under orders of any government, public, municipal or local authority;

 .2 disappearance or shortage, only if revealed when an inventory is made or is not traceable to an identifiable event;

 .3 an Excepted Risk.

If the Contract is in Northern Ireland two further exclusions apply:

 .4 civil commotion;

 .5 terrorist acts - a terrorist is a member of or somebody acting for an organisation which is proscribed under the Northern Ireland (Emergency Provisions) Act, 1973;

terrorism is violence for political ends
including violence to scare the public
or a section thereof.

Site Materials - all unfixed materials and goods delivered to, placed on or adjacent to the Works and intended for incorporation therein.

Note: No definition of temporary works is given.

The insurance does not provide cover for temporary works, plant and similar items of the Contractor etc; insurance of such items should be effected by the Contractor and Sub-Contractors.

All risks insurance policies are not standard. The Contract lists the risks to be covered by the insurance, but it is probable that the wording of policies will vary (as may the cover offered). It is essential to check that the requisite cover is afforded by the policy (perhaps particularly by scrutiny of exclusion clauses in the policy) prior to actually effecting the insurance. The J.C.T. has checked to ensure that the required cover can be obtained.

22.3.1 The assured (Contractor under Clause 22A; Employer under Clauses 22B or 22C) must ensure that the Joint Names Policy(s) (under the applicable of Clauses 22A.1, 22A.3, 22B.1, 22C.1, 22C.2):

(a) provide for recognition of each NS/C as an insured under the relevant Joint Names Policy, or

(b) include a waiver by the insurers of their rights of subrogation (if any) against any NS/C.

Thus, under alternative (b), the insurer waives any right to sue any NS/C for damage caused by the NS/C and covered by (and thence reimbursed under) the Joint Names Policy.

The recognition or waiver for NS/Cs is in respect of loss or damage to the Works and Site Materials caused by Specified Perils under Clauses 22A, 22B and 22C.2. Where Clause 22C.1 applies, the recognition or waiver applies to loss or damage to the existing structures and relevant contents caused by Specified Perils. Any such recognition or waiver applies up to and including the earlier of:

(i) date of issue of Certificate of Practical Completion of the Sub-Contract Works - NSC/4 or NSC/4a, Clause 14.2

(ii) date of determination of the employment of the Contractor (whether disputed or not) under Clauses 27, 28, or 22.C.4.3 if applicable.

The provisions of Clause 22.3.1 also apply to any Joint Names Policy effected under the Contract where the party who was required to insure has not effected the insurance and the Joint Names Policy has, in consequence, been taken out by the other party (e.g. Employer has effected the insurance under Clause 22A.2).

22.3.2 The provisions of Clause 22.3.1 regarding recognition or waiver apply also to Domestic Sub-Contractors. Such recognition or waiver continues up to and including the earlier of:

(i) the date of issue of any certificate or document which states that the Domestic Sub-Contract Works are practically complete, or

(ii) the date of determination of the Contractor's employment as Clause 22.3.1.

The provisions of Clause 22.3.1 do <u>not</u> apply where the Joint Names Policy is effected under Clauses 22C.1 or 22C.3.

Clause 22A: Erection of new buildings—All Risks Insurance of the Works by the Contractor

22A.1 Contractor to take out and maintain a Joint Names Policy for All Risks Insurance. The minimum scope of cover is denoted in Clause 22.2. The amount of cover required is the full reinstatement value of the Works plus any percentage to cover professional fees as stated in the Appendix.

Subject to Clause 18.1.3, the insurance must be maintained up to and including the earlier of:

(a) the date of issue of the Certificate of Practical Completion, or

(b) the date of determination of the employment of the Contractor under Clause 27 or 28 (whether the validity of such determination is contested or not).

Note: (i) Especially during times of high inflation, it is essential to ensure that the cover is adequate for the full reinstatement value, as required - including Variations etc.

(ii) Where an Extension of Time is awarded, it is essential to ensure that the insurance cover is extended up to and including the new Completion Date.

.2 Employer must approve the insurers with whom the Contractor takes out the Joint Names Policy under Clause

22A.1. Contractor to send to the Architect, for deposit with the Employer:

(a) the Policy,

(b) the premium receipt for the Policy,

(c) any endorsements necessary to maintain the Policy, as required by Clause 22A.1, and

(d) premium receipts for any such endorsements.

If the Contractor defaults in taking out and maintaining the insurance required under Clauses 22A.1 and 22A.2, the Employer may:

(a) effect insurance by a Joint Names Policy against any risks regarding which the Contractor's default occurred,

(b) recover the premium amounts (paid or payable by the Employer to effect the necessary insurance) from the Contractor:

i. by deduction from payments to the Contractor under this Contract, or

ii. as a debt of the Contractor.

Note: Any amounts paid by the Employer to effect insurance under the Contract in the event of the Contractor's failure to effect the required insurance can be recovered from the Contractor either as a debt (so, ultimately, by litigation) or by set-off etc. against monies due from the Employer to the Contractor under this Contract; under JCT '80 such sums cannot be transferred from one project between the parties to another project. The usual means of recovering such sums is by contra-charge or set-off against monies due under Interim Certificates issued subsequently; set-off is a common law right.

22A.3.1 If the Contractor independently maintains a Contractor's all risks insurance, such insurance will discharge the Contractor's obligations to insure under Clause 22A.1 provided:

(a) the scope of the cover is at least that required by Clause 22.2,

(b) the amount of cover is adequate for the

Work's full reinstatement value plus any percentage for professional fees (as Clause 22A.1), and

(c) it is a Joint Names Policy in respect of the Works.

The annual renewal date of such Contractor's all risks policy must be supplied by the Contractor and stated in the Appendix.

The Contractor is not required to deposit such policy and premium receipts with the Employer provided that, when reasonably required to do so by the Employer, the Contractor can send documentary evidence that the policy is being maintained to the Architect for inspection by the Employer. Further, such inspection arrangements can also be invoked by the Employer on any occasion in respect of the policy and premium receipts.

.2 The provisions of Clause 22A.2 (set-off etc.) apply in respect of the Employer's rights to insure should the Contractor default in respect of taking out and maintaining the insurance required under Clause 22A.3.1 (contractor's annual all risks insurance).

22A.4.1 If any loss or damage which is covered by the insurance under Clauses 22A.1, 22A.2 or 22A.3 occurs, immediately upon discovering the loss or damage the Contractor must give written notice to the Employer and to the Architect stating the nature, extent and location of such loss or damage.

.2 The occurrence of such loss or damage must be ignored in calculating any amounts payable to the Contractor under or due to the Contract.

.3 The Contractor must reinstate work damaged, replace or repair Site Materials which have been lost or damaged, remove and dispose of any debris and proceed with the proper execution and completion of the Works once any inspection required by the insurers regarding a claim made under the Joint Names Policy under Clauses 22A.1, 22A.2 or 22A.3 has been completed. Again, the Contractor is required to work diligently to effect the repairs etc.

.4 The Contractor must authorise the insurers to pay to the Employer all monies from the insurance under Clauses 22A.1, 22A.2 or 22A.3 regarding loss or damage. The authority is in respect of the Contractor and all those Domestic Sub-Contractors and NSCs who, under Clause 22.3, are recognised as insured under the Joint Names Policy.

.5 The only monies which the Contractor is entitled to receive in respect of the repairs etc. are those paid under the insurance. Thus, the Contractor must bear any costs incurred in excess of those monies paid under the insurance in the execution of the repairs etc. (Any Variation, however, would be valued by the appropriate method and the Contract Sum adjusted accordingly.)

Clause 22B: Erection of new buildings—All Risks Insurance of the Works by the Employer

22B.1 The Employer is to effect the Joint Names Policy for All Risks Insurance as described in Clause 22A.1.

22B.2 (Not Local Authorities Form). As and when reasonably required to do so by the Contractor, the Employer must produce documentary evidence and receipts showing that the Joint Names Policy under Clause 22B.1 has been taken out and is being maintained. (The Contractor will probably desire to inspect the policy, any endorsements and the receipts for premiums paid.)

If the Employer defaults in effecting the requisite insurance, the Contractor may affect the necessary cover and any amounts paid or payable by the Contractor for such premiums must be added to the Contract sum.

As the Contractor is an insured under the Joint Names Policy effected by the Employer, it is not necessary for the Contractor to obtain separate insurance for the risks covered by the Joint Names Policy. Any additional cover required by the Contractor (for hutting, plant etc.) must be effected by the Contractor separately.

22B.3.1 As Clause 22A.4.1.

22B.3.2 As Clause 22A.4.2.

22B.3.3 As Clause 22A.4.3.

22B.3.4 As Clause 22B.4.4.

22B.3.5 Restoration, replacement etc. work executed by the Contractor - paid for as a Variation required by an A.I. under Clause 12.2.

Clause 22C: Insurance of existing structures—Insurance of Works in or extensions to existing structures

22C.1 The Employer must take out and maintain a Joint Names Policy for:

(a) the existing structures,

(b) any relevant part under Clause 18.1.3 from the relevant date, and

(c) contents of such structures owned by or the responsibility of the Employer for their full cost of reinstatement, repair or replacement caused by one or more of the Specified Perils up to and including the earlier of:

i. the date of issue of the Certificate of Practical Completion, or

ii. determination of the employment of the Contractor under Clause 22C.4.3, 27 or 28 (whether validity contested or not).

For the Contractor and all NS/Cs who in respect of Clause 22.3.1 are recognised as insured under the Joint Names Policy under Clauses 22C.1 or 22C.3, the Contractor must authorise the insurers to pay to the Employer all monies from that insurance in respect of any loss or damage.

22C.2 As Clause 22A.1 but includes possible determination of the employment of the Contractor under Clause 22C.4.3.

22C.3 (Not Local Authorities Form.)

As and when reasonably required to do so by the Contractor, the Employer must produce documentary evidence and receipts showing that the Joint Names Policy under Clause 22C.1 or 22C.2 has been taken out and is being maintained.

If the Employer defaults in effecting the requisite insurance under Clause 22C.1, the Contractor may effect the necessary cover for which purpose the Contractor has right of entry and inspection to make a survey and inventory of the existing structures and the relevant contents.

If the Employer defaults in effecting the requisite insurance under Clause 22C.2, the Contractor may effect the necessary cover. (No entry provisions as for default under Clause 22C.1, are required as this insurance is in respect of the Works, which are in the possession of the Contractor.)

Amounts in respect of premiums paid by the Contractor under Clause 22C.3 must be added to the Contract Sum.

22C.4 If any loss or damage which is covered by the insurance under Clauses 22C.2 or 22C.3 occurs, immediately upon discovering the loss or damage the Contractor must give written notice to the Employer and to the Architect stating the extent, nature and location of such loss or damage, and:

.1 The occurrence of such loss etc. must be ignored in any amounts payable to the Contractor under or due to the Contract,

.2 The Contractor must authorise the insurers to pay to the Employer all monies from the insurance under Clause 22C.2 or 22C.3, regarding loss etc. as Clause 22C.4. Such authority is in respect of the Contractor, and all those Domestic Sub-Contractors and NSCs who, under Clause 22.3, are recognised as insured under the Joint Names Policy.

.3 .1 Within 28 days of the occurrence of a Clause 22C.4 loss:

(a) if just and equitable, the employment

of the Contractor may be determined at the option of either party-notice by registered post or recorded delivery,

(b) within 7 days from receipt of a determination notice, either party may give the other a written request to concur in the appointment of an Arbitrator (under Article 5) to decide if such determination will be just and equitable.

.2 In the event of determination of the Contractor's employment, Clause 28.2 (except Clause 28.2.2.6) is to be followed.

.4 If determination of the Contractor's employment is inapplicable:

.1 after any inspection required by the insurers is completed, the Contractor must make good and complete the Works, including replacement etc. of damaged or lost Site Materials; and

.2 such making good, debris disposal etc. is regarded as a Variation required by an A.I. under Clause 13.2.

Clause 22D: Insurance for Employer's loss of liquidated damages—clause 25.4.3

This Clause provides an option for the Employer to insure against loss of liquidated damages where the Architect awards the Contractor an extension of time for loss or damage caused by the occurrence of one or more of the Specified Perils (flood, fire etc.).

22D.1 The Appendix will state either:

(a) Clause 22D insurance may be required, or

(b) Clause 22D insurance is not required by the Employer.

If the Appendix states that Clause 22D insurance may be required, as soon as the Employer and the Contractor enter into the Contract, the Architect must inform the Contractor that either:

(i) the insurance is not required, or

(ii) the Contractor must obtain a quotation for the insurance.

The quotation must be for insurance:

(a) on an agreed value basis (to avoid subsequent disputes over sum(s) payable - the insurers will satisfy themselves that the liquidated damages stated in the Contract are reasonable),

(b) which provides for payment to the Employer of a sum calculated in accordance with Clause 22D.3.

The sum paid to the Employer is to compensate him for his loss of right to claim liquidated damages from the Contractor where the Architect has awarded an extension of time under Clause 25.3 due to loss/damage to the Works etc. (including temporary buildings, plant etc.) which was caused by the occurrence of one or more of the Specified Perils (flood etc.), a relevant Event under Clause 25.4.3.

If the Contractor reasonably requires any additional information to obtain the quotation, the Architect must get it from the Employer.

Once the Contractor has obtained a quotation:

(a) the Contractor must send it to the Architect, as soon as practicable, and

(b) the Architect must then promptly instruct the Contractor whether or not the Employer wishes to accept that quotation.

If the Contractor receives an A.I. that the quotation is to be accepted, the Contractor must:

(i) forthwith take out and maintain the policy until the date of Practical Completion, and

(ii) send the Policy, premium receipt plus any endorsements and their premium receipts, to the Architect for deposit with the Employer.

.2 The sum insured is:

the rate of liquidated damages (as stated in the Appendix) for the period stated in the Appendix.

.3 Payment under the insurance is:

the rate stated in Clause 22D.2 (revised by application of Clause 18.1.4 - partial possession by the Employer) multiplied by the shorter period of either:

(a) that stated in the Appendix, or

(b) that extension of time awarded by the Architect as referred to in Clause 22D.1 (Relevant Event is the Specified Peril(s)).

.4 Amounts spent by the Contractor in effecting Clause 22D insurance are added to the Contract Sum.

If the Contractor defaults in effecting the requisite Clause 22D insurance, the Employer may effect such insurance.

Clauses 22E–K: Model Clauses for use where the Employer does not require Works insurance under 1986 Clause 22A or Clause 22B or Clause 22C

The model clauses, 22E, 22F, 22G, 22H, 22J and 22K are set out in appendix C to Practice Note 22. It is envisaged that only rarely will these clauses be used by private sector Employers but, as Local Authorities usually do not take out insurance, these alternative clauses may be used quite extensively by Local Authority Employers.

The model clauses are alternatives to Clauses 22A, 22B or 22C.2.3 and .4 and require the Employer to assume the risk otherwise subject of the All Risks Insurance in respect of work executed and Site Materials; for work on existing structures the risks assumed by the Employer include those defined as Specified Perils.

If the Employer assumes the All Risks Insurance risks or the Specified Perils risks, the Employer may do so as accepting either:

(a) the risk, or

(b) the sole risk.

Should the Employer choose to accept the risk (rather than the sole risk), the Employer has the right not to pay the Contractor for making good etc. of loss or damage caused by the occurrence of risk assumed by the Employer to the extent that the loss or damage was caused by negligence of the Contractor. Further, if loss or damage were caused to existing structures/ contents through the occurrence of a Specified Peril which was caused by negligence of the Contractor, the Employer could claim against the Contractor for such loss or damage to the extent that it was caused by the Contractor's negligence.

Thus, if the Employer elects to assume the risk, rather than the sole risk, the Contractor would be well advised to effect insurance in respect of the Contractor's potential liability in negligence in respect of those risks assumed by the Employer.

Clause 23: Date of Possession, completion and postponement

23.1 Possession of the site to be given to the Contractor on the Date of Possession (as Appendix). He must regularly and diligently proceed with the execution of the Works and complete (Practical Completion) on or before the Completion Date (as Appendix - Clause 1.3).

Failure by the Employer to give the Contractor possession would be a breach of Contract, the Contractor would thereby be entitled to damages and time for completion of the Contract, if applicable, would be 'at large'. The Architect is not entitled to alter the Date of Possession.

Following the ruling in Trollope & Colls Ltd. v. N. W. Metropolitan Hospital Board (1973), it may be preferable to state the Completion Date as after a time period commencing on the Date of Possession.

23.2 The Architect may issue A.I.s covering postponement of any work to be executed under the Contract.

Such A.I.s may:

(a) provide grounds for an Extension of Time (Clause 25.4.3.1)
(b) provide grounds for a direct loss/expense claim (Clause 26.2.5)
(c) provide grounds for the Contractor to determine his employment (Clause 28.1.3.4)

L. B. of Hounslow v. Twickenham Garden Developments Ltd. (1970). Megarry J.:

(a) 'The Contract necessarily requires the building owner to give the Contractor such possession, occupation or use as is necessary to enable him to perform the Contract.'

(b) Problem of interpretation of 'regularly and diligently' - use of programme to aid interpretation?

.3.1 For purposes of the Works insurances, the Contractor possesses the site and the Works up to and including the date of issue of the Certificate of Practical Completion.

Subject to Clause 18 (partial possession by the Employer), the Employer is not entitled to take possession of any part of the Works until the date of issue of the Certificate of Practical Completion.

Thus, once the Employer has taken possession of the Works (or a relevant part), from the date of issue of the Certificate of Practical Completion, it is the Employer's responsibility to effect any insurance of the Works (or relevant part).

.2 Despite Clause 23.3.1, prior to the issue of the Certificate of Practical Completion, the Employer may use or occupy part or all of the site and the Works provided the Contractor has consented in writing to such occupation or use.

The Employer must notify the insurers under that applicable of Clauses 22A, 22B or 22C.2, .3 and .4 and obtain those insurer's confirmation that the intended use or occupation by the Employer will not prejudice (affect detrimentally) the insurance; this must occur before the Contractor consents to the Employer's use or occupation.

Once the insurer's confirmation has been obtained, the Contractor's consent may not be withheld unreasonably i.e. the Contractor must be able to justify not giving consent to the desired occupation or use by the Employer.

.3 If either Clause 22A.2 or 22B.3 applies and the insurers have required payment of an extra premium in return for their confirmation under Clause 23.3.2, the Contractor must notify the Employer of the amount of that extra premium.

If the Employer still desires to occupy or use the site or Works:

(a) the extra premium must be added to the Contract Sum, and

(b) the Contractor must provide the Employer with the receipt for that extra premium, if requested to do so by the Employer.

Clause 24: Damages for non-completion

See also Practice Note 16.

24.1 If the Contractor fails to complete the Works (Practical Completion) by the Completion Date, the Architect must so certify. This is, of course, subject to Extension of Time awards.

24.2 .1 Provided the Architect has issued a Clause 24.1 certificate, the Contractor must pay or allow to the Employer the whole or part of a sum specified by the Employer in writing and calculated for :

(a) a period between the date of the Clause 24.1 Certificate and Practical Completion

(b) this period at the appropriate rate of liquidated damages (Appendix). (No proof of actual damage is required.)

This is to be before the date of the Final Certificate. Thus the calculation is:

(a) (usually z weeks) multiplied by (b) (usually £x per week) = amount payable.

<u>Note</u>: Both Partial Possession by the Employer and/or Extensions of Time will mitigate the Contractor's liability.

The Employer may recover such sums from any payments to the Contractor under this Contract only, not any other contract between the same parties.

.2 If an Extension of Time is awarded after a liquidated damages amount has been deducted, then there is provision for the Employer to repay such amount to the Contractor but without any interest thereon.

Note: Penalties for default are not recognised in English law. Thus any sum stated as liquidated and ascertained damages must be a reasonable (genuine pre-estimate) estimate of the damage that the Employer would suffer if the project were to be completed late - see Hadley v. Baxendale (1854).

Dunlop Pneumatic Tyre Co. Ltd. v. New Garage & Motor Co. Ltd. (1915) - provides guidance as to what will constitute a penalty (and thereby be unenforceable but to which severance might be applied under Equity).

Peak Construction (Liverpool) Ltd. v. McKinney Foundation Ltd. (1971): If the Employer is in any way responsible for the Contractor's failure to meet the Completion Date, he cannot sue for liquidated damages.

He may still have a claim in respect of the Contractor's contribution to the late completion achieved.

Ramac Construction Co. Ltd. v. J. E. Lesser (Properties) Ltd. (1975) (High Court) - the failure of an Architect to issue a certificate *re* Clause 24.1 will not preclude a dispute regarding the validity of an Employer's counter-claim from going to Arbitration.

If any liquidated damages due to the Employer are not paid or otherwise allowed by the Contractor prior to the issue of the Final Certificate, the Employer's right to recover such amount from the Contractor is, it is submitted, terminated by the issue of the Financial Certificate. Such liquidated damages not recovered cannot subsequently be set off against any claims submitted by the Contractor

This is an area of potential danger for the Architect - such a situation could lay the Architect open to a claim by the Employer for the liquidated damages he can no longer recover from the Contractor. Architects would be well advised to ensure the Employer has obtained all appropriate settlements, including liquidated damages, prior to issuing the Final Certificate.

The Employer maintains the right to recover liquidated damages by ensuring the claim is specified to the Contractor in writing as per Clause 24.2.1 and is agreed and acknowledged by the Contractor, preferably in writing, prior to the issue of the Final Certificate, such notice stating that the Employer may recover the appropriate sum at a later time.

Clause 25: Extension of Time

See also Practice Note 16.

This Clause is closely associated with Clause 26, Clause 25 being time and Clause 26 being money.

A Clause 25 claim may arise on its own but a Clause 26 claim will be associated with a claim under Clause 25.

A successful Clause 25 claim obviates the liability for the Contractor to pay liquidated and ascertained damages for non-completion in respect of the time period for which the Extension of Time is granted.

25.1 Delay includes further delay for this Clause.

25.2 .1.1 Once it is reasonably apparent that the progress of the Works is being or is likely to be delayed, the Contractor must inform the Architect in writing stating the circumstances causing the delay and noting the Relevant Event(s) (see Clause 25.4) he considers applicable.

i.e. The Contractor must indicate the cause of the delay as denoted by Clause 25.4 for any claim to be considered. If the claim is not covered by any of the reasons given by Clause 25.4, it will *not* be valid for consideration of an Extension of Time.

.2 If the Contractor's notice under Clause 25.2.1.1 refers in any way to an NS/C the Contractor must give that

NS/C a copy of the notice.

.2 The Contractor must, about each Relevant Event specified by him in a notice, either within the notice or as soon as possible in writing specify:

.1 the expected effects

.2 the period of expected delay in completing (if any) - also to any NS/Cs named in Clause 25.2.1.2.

(New provision, generally considered most inadvisable under the 1963 Edition.)

.3 The Contractor to give further written notices (including to NS/Cs as named) as necessary or requested by the Architect to keep up to date particulars of delays (Clause 25.2.2.1, and time involved estimates (Clause 25.2.2.2) including any material changes to them.

25.3 .1 On receipt of information as required regarding alleged delays, if in the Architect's opinion:

.1 any of the events stated by the Contractor are Relevant

and

.2 the completion of the Works is likely to be delayed beyond the Completion Date due to these Relevant Events, then the Architect shall give the Contractor a written Extension of Time by fixing such later Completion Date as he considers to be fair and reasonable.

In fixing such later Completion Date, the Architect must state the points covered by the following two sub-clauses:

.3 which Relevant Events he has taken into account

and

.4 the extent to which, if any, he has taken into account omission Variation issued subsequent to the last fixing of the Completion Date.

The Architect shall, if practicable and if he has sufficient information, fix the new Completion Date within 12 weeks, i.e. not later than 12 weeks from receipt of the Contractor's notice.

If the period from the Architect's receipt of the Contractor's notice and the existing Completion Date is less than 12 weeks, he shall fix the revised Completion Date before the existing Completion Date is reached.

Note: It is the Architect's opinion regarding the Relevant Events' applicability and time involved which governs the award of any Extension of Time. The Contractor's written notice is a pre-requisite for an extension as is (now) the provision of his estimate of the delay involved. The Architect does not have to apportion any Extension of Time awarded between the Relevant Events causing the delay, except as required by Clause 26.3 to assist in the calculation of loss/expense awards under Clause 26.1

(1963 Edition - no period to be given by Contractor. Architects usually delayed awards until Practical Completion and then made awards retrospectively (except for strikes). The new system seems preferable.)

Amalgamated Building Contractors Ltd. v. Waltham Holy Cross U.D.C. (1952): no longer applicable but applied to 1963 Edition.

See Miller v. L.C.C. (1934):

(a) If the Architect fails to grant an Extension of Time when he should have done so, liquidated damages are *not* recoverable.

(b) It appears that if the cause(s) of delay is within the control of the Employer or Architect and no applicable Extension has been granted by the Completion Date last determined, the Completion Date is 'at large', i.e. the Contractor must complete within a reasonable time.

(c) If the Contractor disagrees with any Extension award (e.g. no Extension) the Contractor may invoke Arbitration prior to Practical Completion (or alleged Practical Completion) - Article 5.

.2 The first award by the Architect must be an extension of time, i.e. fix a Completion Date later than that originally specified in the Contract.

Subsequent consideration and awards may lead to the Completion Date being revised to one earlier than an extended date due to omission Variations being taken into account in the date revision, provided the relevant omissions were due to A.I.s issued after the last revision of the Completion Date.

Despite omissions, the *Architect* cannot revise the Completion Date to one earlier than that specified in the original Contract - expressed by Clause 25.3.6.

.3 Within 12 weeks from the date of Practical Completion, the Architect must, in writing to the Contractor, do one of the following:

.1 revise the existing Completion Date to a later one if fair and reasonable so to do in the light of Relevant Events.
(Note actual wording - permits review by the Architect.)

.2 revise the existing Completion Date to an earlier one - if reasonable due to omissions subsequent to last fixing of the Completion Date.

.3 confirm the existing Completion Date.

This provision permits the Architect to make an Extension of Time award in the absence of the Contractor's written notice, which therefore in this instance is *not* a prerequisite to an Extension,

.4 provided

.1 the Contractor uses his best endeavours to prevent delays.

This last point often (1963 Edition) has been interpreted to mean that the Contractor must try and make up the time of delays. This is *not* so and accounts for the new 12 weeks award rule being seen as of major importance.

.2 the Contractor must do all that is reasonably required to the satisfaction of the Architect to proceed with the Works.

.5 The Architect to notify every NS/C of each revision to the Completion Date.

.6 The Architect may not revise the Completion Date to one earlier than that stated in the Appendix.

25.4 List of Relevant Events:

.1 *Force majeure* - Act of God; man-made events beyond the control of the parties.

.2 *Exceptionally* adverse weather conditions - normal adverse weather (predictable) is assumed to be incorporated into the programme - note effects of location, time of year and stage of job - presumably now covers excessive heat and drought as well as cold, rain, snow and frost.

Note: site weather records - diary and use of Meteorological Office data to establish validity of claims. It is often good practice to record delays due to weather and actual weather conditions on a daily basis and to agree such records at the monthly site meeting.

Walter Lawrence and Son Ltd. v. Commercial Union Properties (UK) Ltd. (1984): held that the effects of exceptionally adverse weather (under JCT 63, so exceptionally inclement weather was considered) should be assessed regarding the time at which the works affected are executed, rather than at the time indicated on the programme (if any) for their execution.

.3 Loss or damage from 'specified Perils' (flood, etc.).

Note: despite the requirements of Clause 22, separate notice is also required under this Clause.

.4 Civil commotion, strike or lock-out - at the site, Sub-Contractor's or Supplier's premises, or transport directly connected with the work execution.

.5 Compliance with A.I.s:

.1 discrepancies, divergences (Clause 2.3)
Variations (Clause 13.2)
provisional sums (Clause 13.3)
postponement (Clause 23.2)
antiquities (Clause 34)
NS/Cs (Clause 35)
NSups (Clause 36

.2 inspection and testing where items *do* comply with the Contract (Clause 8.3).

.6 Non-receipt by the Contractor in 'due time' (taken to be time commensurate with allowing the Contractor to receive, process the information and set up and execute the work in accordance with normal practice. Thus, he should receive the information in a reasonable time prior to the time at which the items of work are scheduled to be done).

'Necessary instructions, drawings, details or levels from the Architect for which he (the Contractor) specifically applied in writing.'

The provision of the programme is of significance, especially if together with a schedule of key dates for the release of information (*not* a Contract Document), in a practical sense.

Contractually the specific written application for the required information is essential and so even a master programme denoting key dates is *insufficient*,

'provided that such application was made on a date which having regard to the Completion Date was neither unreasonably distant from nor unreasonably close to the date on which it was necessary for him to receive the same'.

This situation is concerned with Extension of Time. The obligation upon the Architect to provide the necessary information at the requisite time is now greater due to the requirement of the programme provision by the Contractor but the Contractor must still properly apply, in writing, for the necessary information.

.7 Delay by NS/Cs or NSups which the Contractor has taken all practicable steps to avoid or reduce.

Note: Contractor is obliged to minimise the delays as far as possible, not to make up time.

City of Westminster v. J. Jarvis & Sons Ltd. and Peter Lind Ltd. (1969): the question concerning the Completion Date when latent defects were discovered in piles - S/C was in breach; not delay as Practical Completion of the piling S/C had been achieved.

Trollope & Colls Ltd. v. N. W. Metropolitan Hospital Board (1973): Completion Date on a phased project. If actual dates given these hold good despite delays to previous phases of the project.

Failure of the Employer to give the Contractor possesion of the site is not a ground for an Extension of Time. In such circumstances the time for completion will be 'at large' with the consequent effect upon the possibility of deducting liquidated damages. (Any attempt to require the Contractor to complete by the original Completion Date would necessitate the Contractor's working faster than he intended and would thereby change the terms of the offer (Tender) and its acceptance and is therefore invalid.)

.8.1 Delay or failure to execute work by Artists and Tradesmen. These are direct to the Employer and thus outside the scope of the Contract, as such, and the Contractor's control.

It is probable that the Employer could recover damages from the Artists and Tradesmen if they are independent contractors.

.2 Delay in or failure to supply goods and materials which the Employer is contractually obliged to supply.

.9 By *the Government of the United Kingdom* , exercising any statutory power after the Date of Tender and thereby adversely affecting the Contractor's procuring necessary labour, goods, fuel and energy essential to the proper execution of the Works.

.10.1 The Contractor's inability to secure such labour as is *essential* to the proper carrying out of the Works.

This must be:

(a) beyond the Contractor's control, and

(b) not reasonably foreseeable at the Date of Tender.

.2 Reproduction of Clause 25.4.10.1 but for goods and materials.

In such instances the Contractor must exercise reasonable foresight - this will be applicable to his ordering and thereby pass on any liability to S/Cs and Suppliers. He must endeavour to procure the requisite specified items as far as possible.

The items must be essential to the Works - alternatives must be considered as must alternative methods of working and types of construction.

.11 Execution delay or failure by a Statutory Undertaker in

pursuance of a statutory obligation.

This does not cover work done outside the scope of a statutory obligation. In such a case the Statutory Undertaker would be acting as an ordinary Nominated or Domestic S/C. If acting as an NS/C the delay, etc., is covered by Clause 25.4.5.1, if acting as a Domestic S/C the delay must be borne by the Contractor.

.12 Failure by the Employer to give, in due time, access to the site via property which he possesses and controls in accordance with the Contract Bills and/or Drawings, provided any required notice for such access has been given by the Contractor to the Architect, or the Architect and Contractor have themselves agreed an access provision (via the Employer's property). Access includes both ingress to and egress from the site - perhaps particularly relevant to conversion projects.

Following Porter v. Tottenham U. D. C. (1915), the Contractor assumed the risks of access obstructions caused by third parties. Due to the wording of this Clause, the ruling would appear still to be applicable - 'in the possession and control of the Employer', is the vital issue.

Following Hounslow Borough Council v. Twickenham Garden Developments Ltd. (1971), in considering an extension of time, the Architect is entitled to take into account any amount by which the Contractor is ahead of programme and to reduce any extension accordingly.

Clause 26: Loss and expense caused by matters materially affecting regular progress of the Works

This Clause is generally viewed as the money element of the Clauses 25 and 26 combination. As such, the items covered whereby the disturbance of the regular progress of the Works is material and causes direct loss and/or expense to the Contractor are less in number and scope than the Relevant Events specified in Clause 25.

The matters listed by Clause 26 as being a basis for a claim are limited to those Relevant Events considered to be within the control of the Employer and his agents (e.g. the Architect).

26.1 The Contractor's making written application to the Architect is a condition precendent for a claim under this Clause. The written application must state:

he has incurred or believes he will incur direct loss and/or expense in the execution of the Contract.

Note: It must not be possible under the Contract for the Contractor to be reimbursed under any other provision as the loss is (or will be) attributable solely to the disturbance of the regular progress of the Works, that the progress is *materially* affected and that this has been caused by one or more of the matters listed in this Clause (Clause 26.2).

.1 The Contractor's application must be made as soon as the delay is apparent, is likely or should have reasonably become apparent. It may affect the entire Works or any part thereof.

F. G. Minter Ltd. v. Welsh Health Technical Services Organisation (1980) (Court of Appeal):

'In the building and construction industry "cash flow" is vital to the Contractor and delay in paying him for the work he does naturally results in the ordinary course of things in his being short of working capital, having to borrow capital to pay wages and hire charges and locking up in plant, labour and materials capital which would have been invested elsewhere.

The case was concerned with the 1963 Edition of the J.C.T. Contract (with certain amendments), the major point at issue being the inclusion of finance charges in a claim for direct loss and/or expense (Clause 24 of the 1963 Edition), which, it was decided, would be applied to the period from the contractor's incurring the loss/expense to the giving of the loss/expense notice. Under JCT 80 a contractor may give notice about future anticipated loss/expense, under Clause 26.1, and so, it is submitted, interest may apply for the incurrence of the loss/expense to the point of settling the claim; probably certification of the appropriate sum, that being the juncture at which a debt is created. Interest may continue to accrue after practical completion.

The result of the appeal hearing contains two points of relevance to any claims under Clause 26 of the 1980 Edition which would, by implication, apply:

(a) Direct loss and/or expense may include finance charges or interest,

and

(b) It is (probably) open to the Contractor to make a single application for the reimbursement of the capital sum together with the interest thereon from the date the expenditure was incurred until the date of certification. (This is because the Contractor's written application under Clause 26.1 must state that he has incurred *or is likely to incur* direct loss and/or expense in the execution of the Contract ...) However, he is probably safer to make a series of claims in respect of a continuing loss/expense, e.g. finance charges.

In Rees & Kirby Ltd. v. Swansea City Council (1985), the Court of Appeal took the view that finance charges should be calculated on the same basis as a bank overdraft or deposit interest in that interest should be compounded at quarterly intervals.

Such finance charges must be distinguished from interest on a quantified debt which is paid late, as this interest is payable only where contract terms so permit, otherwise it is not payable under common law.

Following Croudace Ltd. v. L. B. Lambeth (1985), failure of the Architect to include finance charges when settling contractor's claims is a breach of contract rendering the Employer liable.

Note: The Contractor is obviously regarded as a construction expert (professional) in the context of foreseeing delays and the causes thereof.

.2 The Contractor must supply the Architect with any information in support of the application to allow the Architect to decide its validity and effects. This is qualified by:

(a) reasonableness

(b) the request for information by the Architect.

.3 The Contractor must submit details of loss and/or expense to the Architect or Quantity Surveyor to enable that party to ascertain the extent of the loss. This again is qualified by:

(a) reasonableness

(b) the request of the Architect/Q.S. for the details.

As soon as the Architect is of the opinion that the Contractor's application has some validity he (or the Q.S. under his instruction) must from time to time ascertain the amount of such loss and/or expense incurred by the Contractor.

It should be noted at this juncture that this Clause does *not* cover the more general cases where circumstances have changed thereby increasing the cost of executing the Works.

26.2 The matters giving rise to a claim are specified:

.1 The non-supply of information by the Architect, the Contractor having specifically requested the informa-

tion in writing ......
(Reproduction of Clause 25.4.6).

See Trollope & Colls Ltd. v. Singer (1913).

.2 Opening up and testing where the items are found to be in accordance with the Contract.
(Clause 25.4.2).

.3 Discrepancies between Contract Drawings and Bills.
(Clause 25.4.5.1).

.4.1 Delays due to Artists and Tradesmen.
(Reproduction of Clause 25.4.8.1).

.2 Supply failure by the Employer.
(Reproduction of Clause 25.4.8.2).

.5 A.I.s *re* postponement.
(Clause 25.4.1).

.6 Failure of the Employer to give access.
(Reproduction of Clause 25.4.12).

.7 A.I.s for Variations and expenditure against provisional sums.
(Clause 25.4.5.1).

26.3 The Architect is to state in writing to the Contractor his Extension of Time award regarding certain specified Relevant Events to the extent that this is necessary to determine the amount of any direct loss and/or expense.

(*re* Clauses 2.3; 13.2; 13.3; 23.2; 25.4.5.2; 25.4.6; 25.4.8; 25.4.12.)

26.4 .1 Provides for the Contractor to pass on to the Architect any written application, properly executed under Clause 13.1 of NSC/4 or 4a, of a Nominated S/C for direct loss/expense due to delays.

The Architect, if satisfied about the delays and causes (or the Q.S. on his behalf), shall ascertain the amount of the loss/expense.

.2 If necessary to do so to determine the amount of such loss, the Architect must inform, in writing, both the

Contractor and NS/C of any relevant Extension of Time in respect of certain specified Relevant Events.

(S/Contract Clauses 11.2.5.5.1 (in reference to main Contract 2.3; 13.2; 13.3; 23.2) and 11.2.5.5.2; 11.2.5.6; 11.2.5.8; 11.2.5.12.)

26.5 Any awards to be added to the Contract Sum as awarded - see also Clause 3.

26.6 The provisions of Clause 26 are without prejudice (i.e. have no effect upon) any other rights and remedies the Contractor may possess.

(e.g. claim for breach by the Architect's non-supply of requisite information - see Trollope & Colls Ltd. v. Singer (1913).)

In International Minerals and Chemical Corporation v. Karl O. Helm AG and Another (1986) "... the surviving principle of legal policy is that it is a legal presumption that in the ordinary course of things a person does not suffer any loss by reason of the late payment of money. This is an artificial presumption but is justified by the fact that the usual loss is an interest loss and that compensation for this has been provided for and limited by statute. It follows that a plaintiff, where he is seeking to recover damages for the late payment of money, must prove not only that he suffered the alleged additional special loss and that it was caused by the defendant's default but also that the defendant had knowledge of the facts or circumstances which make such a loss a not unlikely consequence of such a default. In the eyes of the law, those facts or circumstances are deemed to be special ..."

Special damages occur under the second branch of the rule in Hadley v. Baxendale (1854) as being within the knowledge (reasonable expectation) of the defendant party at the time of contracting.

As cash flow is acknowledged as being vital in the construction industry and delay in making a payment quite obviously would involve the Contractor's incurring additional finance charges (interest), such additional costs should be recoverable.

Clause 27: Determination by Employer

27.1 The Employer's rights under this Clause are expressly without prejudice to his other possible rights and remedies in the event of default by the Contractor's:

.1 *wholly* suspending execution of the Works prior to completion (practical) without reasonabl e cause
i.e. the Contractor may suspend execution of part of the Works but not the entire project

.2 failure to proceed regularly and diligently with the Works

.3 refusal or persistent failure to remove defective items which *materially* affect the Works and regarding which the Architect has given him written notice to remove

.4 failure to comply with Clause 19 or 19A*.
(Assignment and Sub-Contracts; Fair Wages*)

*19A is applicable only to the Local Authorities Form.

The Architect may give written notice specifying the default to the Contractor by registered post or recorded delivery.

Note: The delivery method of the written notice is specifed. Sub-Clauses .1, .2, .3 represent repudiatory breach of Contract thus, for any of these, the Employer has a common law right to *'determine the contract'*, as discussed by the House of Lords in Photo Production Ltd. v. Securicor Transport Ltd. (1980).

If the Contractor

(a) continues the default for 14 days from receipt of the notice, or

(b) repeats the default after receipt of the notice (obviously a separate occasion, not a continuation)

the Employer *may* within 10 days of (a) or (b), by registered post or recorded delivery, serve notice of determination of the Contractor's employment upon the Contractor.

The notice not to be given unreasonably or vexatiously. (The notice would, in such instances, be invalid.)

<u>Note</u>: The determination of employment applies only to that Contract on which the default occurs and the action is taken.

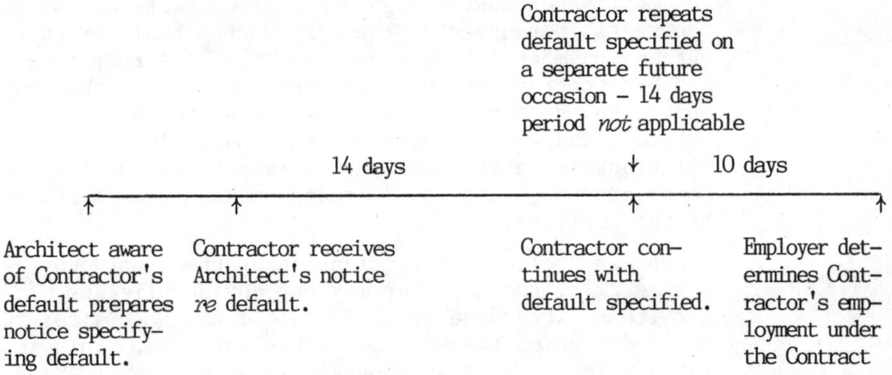

Contractor repeats default specified on a separate future occasion – 14 days period *not* applicable

14 days ↓ 10 days

↑	↑	↑	↑
Architect aware of Contractor's default prepares notice specifying default.	Contractor receives Architect's notice *re* default.	Contractor continues with default specified.	Employer determines Contractor's employment under the Contract

27.2 If the Contractor goes into liquidation except for the purposes of re-construction of the firm or an amalgamation, the employment of the Contractor is automatically determined.

However, if the Employer and liquidator (etc.) of the Contractor so agree, the employment may be reinstated and continued.

27.3* The Employer may determine the Contractor's employment for corruption on the part of the Contractor.

This is a widely-scoped Clause.

*This Clause appears only in the Local Authority Form.

27.4 If the employment of the Contractor is determined under this Clause and not reinstated:

.1 The Employer may employ another Contractor to complete the Works who may use all the items (huts, plant and materials) on or adjacent to the Works and intended for use thereon.

This provision includes the use by the new Contractor of the original Contractor's *hired* plant.

.2.1 Except in the case of liquidation of the Contractor as defined in Clause 27.2, if the Architect or Employer requires, within 14 days of the date of determination, the Contractor must assign all Sub-Contracts and supply Contracts in respect of the Work to the Employer (without any payment for such assignment). The Suppliers and Sub-Contractors who have Contracts so assigned may object to any further assignment (e.g. to the new Contractor) and would probably then remain in direct relationship to the Employer in respect of the new Contractor as Artists and Tradesmen and items supplied by the Employer.

.2 Except in the case of liquidation of the Contractor as Clause 27.2, the Employer may pay any Supplier or S/C for items (work done or goods, etc., *delivered*) for the Works for which the Contractor has not paid. This is in addition to similar provisions in respect of NS/Cs. Any such payments may be contra-charged against the Contractor.

.3 The Architect must inform the Contractor in writing (usually, but not necessarily, and A.I.) when he requires the Contractor to remove all his temporary buildings and plant (including hired items).

If the Contractor does not remove the items within a reasonable time of the Architect's written request, the Employer may:

(a) remove the items in question (he is not liable for any loss or damage thereto under the Contract)

(b) sell the items

(c) hold the proceeds of the sale to the credit of the Contractor after deduction of all costs incurred in the removal and sale.

.4 The Contractor to pay or allow to the Employer any direct loss/damage caused to the Employer by the determination.

Once the Works have been completed (as Clause 27.4.1) - most equitably Final Completion - and the accounts for the Completion have been verified (must be within a reasonable time from the physical completion), the Architect must certify:

(a) amount of expenses properly incurred by the Employer

and

(b) amount of direct loss/damage to the Employer due to the determination.

These accounts, when considered with any sums due to the Contractor, constitute a debt between the parties.

Illustration:

Contract Sum		£800 000
Value of Work executed		£250 000
Certificates (honoured)		£220 000

Thus the value of work to be completed by the original Contractor, A, on the basis of the Contract Sum = £550 000
(Assuming no Variations or Fluctuations.)

Final account of Contractor B in completing the original Contract Work =		£600 000
Excess price of B over A to complete		£ 50 000
Delay costs (A off to B on site)	say	£ 10 000
		£ 60 000
Deduct:		
Sale of A's plant	£ 25 000	
Less sale expenses	£ 5 000	£ 20 000
		£ 40 000
Deduct:		
Owed to Contractor A but not certified prior to determination		£ 30 000
Debt owed by Contractor A to Employer		£ 10 000

i.e. the Employer is Contractor A's creditor for £10 000.

* Determination of Contractor's Employment

Check-list for Action by the Architect

1 .1 Information is received that the Contractor is in financial difficulty/has become insolvent/is in liquidation.

.1 Establish the validity of the information taking care not to spread what may be a false rumour.

.2 Obtain the precise situation from the Receiver or Liquidator, if appointed.

.3 Establish the situation on the site and take the necessary steps to prevent unauthorised removal of items - see Clause 27.4.2.

The actions outlined above should not be undertaken too forcefull until the full facts have been established, i.e. keep a 'low profile'.

.2 If not determined automatically due to liquidation, etc., the Contractor's employment may be determined under Clause 27.1 due to non- or inadequate performance. Such determination is at the option of the Employer.

2 The Employer should be fully advised of the situation and his instructions obtained.

.1 Action should be discussed and recommended regarding:

.1 Liaison with the Receiver.

.2 Safety and security measures (including any already implemented).

.3 Re-insurances.

.4 Completion of the Works - by the same or another Contractor?

.5 Assignment of Sub-Contracts and supply Contracts.

.6 Direct payments to Sub-Contractors and Suppliers.

.2 It should be agreed who is to take any action considered necessary.

3 The other consultants should be informed of the situation as it develops and the agreed action. Particularly, the Quantity Surveyor will be involved.

.1 The items covered in 2 above should be considered.

.2 An accurate record of site conditions, progress, stocks, etc., is vital; photographs and an audit are usually required.

.3 The procedure for completing the project should be agreed and implemented.

4 Written requirements (usually by A.I.s) should be issued to the Contractor. Typically these will include such contractual matters as:

Non-removal of plant, materials and hutting on site which will be required for the completion of the project,
Assignment of Sub-Contracts and Supply Contracts,

and such 'practical matters' as:

Closure of the site,
Safety and security measures,
Safety of site documents and records.

5 It may be necessary for another Contractor to be employed to implement any necessary safety and security measures.

6 The Clerk of Works will have much useful information - this should be fully documented.

If on site, he must be instructed regarding who may have access to the project and what items, if any, are to be delivered, removed and any work to be carried out.

7 If the Contractor is in liquidation, the Liquidator should be consulted to determine his intentions regarding the Contractor's future. It may well be advantageous (if possible) to arrange for the Contractor's employment to be reinstated (Clause 27.2).

8 It may be necessary for the Quantity Surveyor to prepare tender documents for the completion of the project.

9 A full discussion should take place between the consultants and the Employer should be advised of the most appropriate policy to obtain a suitable Contractor to complete the project:

.1 Competitive Tender?

.2 Negotiation?

10 Arrangements should be made to enable the new (or reinstated) Contractor to recommence site working as soon as possible.

11 Assignments of Sub-Contracts and Supply Contracts must be formalised, if required, from the original Contractor to the Employer. Unless the Sub-Contractors or Suppliers object (proviso of reasonableness) the assignment is most usefully effected from the original to the new Contractor. (Clause 27.2). The notice from the Architect requiring assignment must be given within 14 days of the date of determination.

12 When the plant of the original Contractor is no longer required, arrangements must be made for its removal from the site - this may involve sale of the plant by the Employer.

13 Once the project has been completed and the accounts verified (reasonable time from Completion) the final account between the Employer and the original Contractor must be settled. Note the provisions regarding set-off by the Employer in respect of the additional costs incurred by him to achieve completion.

*Reference: Ginnings, A. T. - 'Determination of Employment under the Standard Forms of Contract for Construction Works' - The Quantity Surveyor, February 1978, pp 97-101.

Clause 28: Determination by Contractor

28.1 The contractual provisions are expressly without prejudice to any other rights and remedies of the Contractor.

The Contractor may (registered post or recorded delivery) serve notice of determination of his employment under the Contract to either the Employer or Architect. The employment is immediately determined unless the notice has been given "unreasonably or vexatiously".

J. M. Hill Ltd. v. L. B. Camden (1982): for determination to be unreasonable, it must be totally unfair and, almost, smacking of sharp-practice. The fact that giving one reason for giving the determination notice was the Contractor's wish to protect himself against a potentially crippling loss and that it would result in his getting profit does not make the Contractor's action unreasonable.

The situation, stated in J. M. Hill was expanded and reinforced in J. Jarvis Ltd. v. Rockdale Housing Association (1985): it was acknowledged that a contractor had a right to protect his interests in deciding what actions to take - "...it is only where there is a gross disparity between the benefit to him and the burden to the Employer that the exercise of his right even approaches unreasonableness."

Thus, it appears that the burden of proof, to establish that a contractor has acted 'unreasonably or vexatiously' is a heavy one.

The Contract provides the following causes of valid determination by the Contractor:

.1 The Employer fails to honour a certificate (except due to V.A.T. Agreement) within 14 days of its issue and so continues for 7 days from receipt of a notice from the Contractor (by registered post or recorded delivery).

The notice will state that if payment is not duly made within 7 days from its receipt by the Employer, a notice of determination under Clause 28 will be served.

Mersey Steel & Iron Co. Ltd. v. Naylor Benzon & Co. Ltd. (1984): payment of a previous instalment is not a condition precendent to the right to claim further instalments.

In Canterbury Pipelines Ltd. v. The Christchurch Drainage Board (1979) - New Zealand Case - it was asserted that there is no *common law* right in English building law for a Contractor to suspend work because of the wrongful withholding of a certificate or progress payment.

.2 "The Employer interferes with or obstructs the issue of *any certificate* due under this Contract."

This will include payments, Practical Completion Certificates.

R. B. Burden Ltd. v. Swansea Corporation (1957): The Clause does not include instances where certificates have been issued negligently, and thus incorrectly. The interference or obstruction must be 'active'. A list of such instances would be very lengthy but would include such examples as:

(a) The Employer directing the Architect as to the amount to be certified.

(b) The Employer refusing access to the site for the Architect for the purpose of issuing a certificate.

.3 The execution of the whole, or majority of the uncompleted Works is suspended for a *continuous* period as stated in the Appendix.

The periods suggested by the J.C.T. in a footnote to the Appendix are:

'Specified Perils' (Clause 28.1.3.2) - 3 months
Other reasons under Clause 28.1.3 - 1 month

Note: This does expressly *not* apply to work involved in

making good defects as Clause 17.

The reasons for the suspension of work are:

.1 *Force majeure*

.2 Loss or damage to the Works caused by 'Specified Perils' - (unless caused by the Contractor or those for whom the Contractor is responsible) provided either Clause 22A or Clause 22B applies.

.3 Civil commotion

.4 A.I.s regarding

(a) Clause 2.3 - discrepancies or divergencies
(b) Clause 13.2 - Variations
(c) Clause 23.2 - postponement
(unless due to the Contractor's negligence or default)

.5 Non-receipt by the Contractor of properly requested information in due time for the work execution (See Clause 25.4.6)

.6 Delay by Artists and Tradesmen, or Employer in supplying goods, etc., direct

.7 Opening up for inspection/testing (unless items prove *not* to be in accordance with the Contract).

Note: All the above reasons are also Relevant Events under Clause 25 giving grounds for an Extension of Time claim.

28.2 Upon determination of the Contractor's employment under the Contract, without prejudice to the other rights of the parties and expressly without affecting the rights under Clause 20 (injury to persons and property and Employer's indemnity) which are to continue to apply before and during the removal by the Contractor and Sub-Contractors of their property (under Clause 28.2.1), the rights and liabilities of the parties are:

.1 The Contractor shall, as soon as possible, and with due regard to the safety of persons and property, remove his temporary buildings, plant and goods and allow Sub-Contractors to do the same. This provision is subject to Clause 28.2.2.4 regarding the passing of property to the Employer - items which the Contractor has purchased or is purchasing properly for the Works become the property of the Employer when the Employer pays the Contractor for them, see also Clause 19.4.2.2.

Thus, if the Employer has *not* paid the Contractor for such items, they may properly be removed.

.2 Subject to consideration of previous payments, the Employer must pay the Contractor:

.1 the total value of work completed at the date of determination, calculated as for an Interim Certificate. Such valuation must also take into account:

(a) direct loss/expense due to delays (Clause 26)
(b) direct loss/expense due to antiquities (Clause 34.3)
(c) property in materials/goods (Clause 28.2.2.4)
(d) reasonable cost of removal of Contractor's and Sub-Contractors' items (Clause 28.2.1)
(e) direct loss/damage to a NS/C due to the determination (Clause 28.2.2.6)

.2 the total value of work partly completed at the date of determination, valued in accordance with the rules for valuing Variations and subject to provisions (a) - (e) above

.3 sum for direct loss/expense due to:

(a) delays (Clause 26)
(b) antiquities (Clause 34.3)

.4 cost of materials or goods correctly for the Works for which the Contractor has paid or must pay. On payment therefore by the Employer, the property in those goods and materials passes to him

.5 reasonable cost of removal of Contractor's and Sub-Contractors' items (Clause 28.2.1)

.6 direct loss/damage to the Contractor or any NS/C caused by the determination.

.3 The Employer must inform the Contractor in writing of any 'balancing payments' under Clause 28.2.2 attributable to any NS/C. He must also inform each NS/C in writing.

It may be possible for the Contractor (and Sub-Contractors and Suppliers) to claim loss of profit in respect of a project subject to determination by the Contractor, the loss of profit being on the uncompleted contract work and forming part of the direct loss/expense claim (Clause 28.2.2.6).

Wraight Ltd. v. P.H.T. Holdings Ltd. (1968)

Typical determination time lag diagram:

Valuation with Q.S.	Architect receives Q.S.'s recommend-ation	Architect issues Interim Certificate = debt to Contractor	No payment made. Then Contractor sends notice of determination intention	Employer receives Contractor's notice
↓ 3-4 days	↓ 3-4 days	↓ 14 days	↓ 3 days	↓ 7 days
		period of honouring	↑ 3 days register-ed post or recorded delivery	↑ Still no payment, Contractor serves notice of determination.

Lintest Builders Ltd. v. Roberts (1980): the Contractor is not in breach of his obligation to proceed with the works 'regularly and diligently' merely because some defective work has been done; however, this situation probably is a matter of degree, necessary remedial work (to remedy defects etc.) may be set-off by the Employer against any sums due to the Contractor as under Clause 28.2 there are accrued rights of the Employer to have defective work rectified - such rights arise when the defective work is done.

Clause 29: Works by Employer or persons employed or engaged by Employer

This Clause is most usually termed 'Artists and Tradesmen' (following the terminology, for convenience, of the 1963 Edition).

29.1 Provided the Contract Bills contain sufficient information for the Contractor to execute the Works in the presence of the Employer or others directly engaged by him executing work which is not part of the Contract, the Contractor must permit such Artists and Tradesmen to execute their work on the site.

29.2 If the Contract Bills do not provide such requisite information regarding work of Artists and Tradesmen, the Employer may, but with the Contractor's consent (not unreasonably withheld), arrange for the execution of the work concerned.

Any extra costs thereby caused to the Contractor are recoverable - see Clause 13.1.2 - restrictions imposed by Employer; Clauses 25.4.8.1 and 26.4 - delays and direct loss/expense.

29.3 Every Artist and Tradesman is the responsibility of the Employer, expressly in respect of injury to persons and property and Employer's indemnity (Clause 20).

Clause 30: Certificates and payments

See also Practice Note 18.

30.1 .1.1 The Architect must issue Interim Certificates stating the amount due to the Contractor from the Employer, usually at monthly intervals (see Clause 30.1.3 and Appendix).

Note: The monthly intervals are *calendar months*, not lunar months, thus 12 payments per year (Section 61 of the Law of Property Act, 1925).

The Contractor is entitled to payment of the sum stated due in an Interim Certificate within 14 days from the date of its issue (period of honouring).

Lubenham Fidelities & Investment Co. Ltd. v. South Pembrokeshire D.C. (1985):

(a) the Architect's issuing of a Certificate stating a sum is due to the Contractor from the Employer is a condition precedent to the Contractor's entitlement to be paid that sum,

(b) if either party is dissatisfied about the sum certified, if discussion does not result in a correction being made to the Certificate by the Architect, the remedy is to invoke Arbitration (under Article 5),

(c) even if the Architect has certified an incorrect sum, the Employer must pay the net amount stated as due (subject to any Arbitration, as noted).

Following C. M. Pillings v. Kent Investments (1985), provided the Employer can put forward a reasonable case, he may seek arbitration to decide the validity of an Interim Certificate before being obliged to pay the Contractor.

.2 The Employer may set-off monies due from the Contractor to him under the Contract against any amounts due to the Contractor under an Interim Certificate. This right is qualified by the provisions regarding Retention in respect of direct payments to NS/Cs (set-off against the retention held against the Contractor, as Clause 35.13.5.4.2).

.3 If the Employer exercises his right of set-off under the Contract, he must inform the Contractor, in writing of his reasons for the set off.

Note: This is a contractual provision; the Employer also has rights of set-off at common law,

.2 Interim Valuations:

(a) to be made by the Quantity Surveyor
(b) whenever the Architect considers them to be necessary to determine the amount to be stated due in an Interim Certificate.

Note: If fluctuations are to be recovered under the N.E.D.O. Formulae provisions, Clause 40.2 requires this Clause to be amended to require an Interim Valuation to be made as a pre-requisite for the issue of each Interim Certificate.

Following the ruling in Sutcliffe v. Thackrah (1974) the Architect would be well advised to follow normal procedures and base each Interim Certificate upon a Q.S.'s valuation.

See also Hedley Byrne & Co. Ltd. v. Heller & Ptnrs. (1964).

.3 Interim Certificates must be issued at the intervals (usually calendar months) stated in the Appendix until the Certificate of Practical Completion is issued. Partial possession by the Employer will be relevant in this context.

After the issue of the Certificate of Practical Completion, Interim Certificates may be issued as and when further amounts are ascertained as payable to the Contractor from the Employer. Here the *minimum* period between Interim Certificates is one calendar month.

30.2 The amount stated as due in an Interim Certificate is the gross valuation (Clause 30.2) less the following:

(a) Retention - as Clause 30.4, *and*

(b) The total amount stated as due in previous Interim Certificates (which, normally, will be previous payments).

This provision is subject to any mutual agreement (between the parties) regarding stage payment. Such agreements are very unusual for J.C.T.-governed building projects but are quite common in international work - see F.I.D.I.C. Form.

The gross valuation is calculated from the rules in Clauses 30.2.1 to 30.2.3 (inclusive). The amount so calculated must be as at a date not more than 7 days prior to the date of the Interim Certificate.

The components of the gross valuation are:

.1 These are subject to Retention

.1 'the total value of work properly executed by the Contractor' (usually, but not necessarily, calculated by multiplying the measured or assessed quantities of work properly executed by the appropriate rates given in the B.Q.) including Variations and any applicable formulae adjustments (Clause 13.5 and 40). This also includes items on '*daywork*'.

Only the value of work which has been *properly executed* in accordance with the Contract must be included; see Sutcliffe v. Chippendale & Edmondson (1971) and Townsend v. Stone Toms & Ptnrs. (1985). The Architect must decide whether any work has not been properly executed and so must be excluded. If defective work is suspected (say by the Q.S. performing a valuation), the suspected defective work and the value thereof should be drawn to the attention of the Architect for a decision to be made prior to certification.

.2 the total value of 'materials and goods on site' provided

(a) they are properly for the Works, and

(b) they are not on site prematurely, and
(c) they are adequately protected and stored.

.3 the total value of materials and goods off site, at the discretion of the Architect, subject to the stipulated provisions under Clause 30.3.

.4 the amounts of Nominated Sub-Contractors' items (NS/C Clause 21.4.1), except final payments.

.5 the amount of Contractor's profit (as Contract Bills or agreement, as applicable) on the amounts included in respect of Nominated Sub-Contractors (Clause 30.2.1.4, 30.2.2.5, 30.2.3.2).

Amounts in respect of Nominated Suppliers will be included in materials on or off site or work done and will be subject to additions for Contractor's profit as appropriate.

.2 These are *not* subject to Retention

.1 amounts to be included due to:

(a) Clause 6.2 - fees or charges
(b) Clause 7 - errors in levels and setting out
(c) Clause 8.3 - inspection and testing
(d) Clause 9.2 - royalties
(e) Clause 17.2 - making good of defects
(f) Clause 17.3 - making good of defects
(g) Clause 21.2.3 - executed risks
(h) Clauses 22B and 22C - non-insurance against All Risks or 'Specified Perils' by Employer.

.2 (a) loss/expense due to delays (Clause 26.1)
(b) loss/expense due to Antiquities (Clause 26.1)

.3 final payments to Nominated Sub-Contractors - Clause 35.17

.4 payments to the Contractor in respect of 'traditional' fluctuations provisions (Clause 38 and 39)

.5 Nominated Sub-Contractors' amounts as Clause (NSC/4 or 4a) 21.4.2

.3 *Deductions not* subject to Retention

.1 'traditional' fluctuations as Clause 38 or 39 - amounts to be allowed to the Employer by the Contractor

.2 NSC/4 or 4a Clause 21.4.3 - Nominated Sub-Contractors' 'traditional' fluctuation allowances to the Contractor.

30.3 Materials and goods may be included in Interim Certificates prior to delivery, at the discretion of the Architect. They must be (collectively referred to here as materials):

.1 intended for incorporation in the Works

.2 completed and ready for incorporation

.3 (a) set apart from other stocks (i.e. they are 'ascertained')
(b) visibly marked to identify:
(i) the Employer, or the person under whose order the items are being held
(ii) 'their destination as the Works'.

.4 if ordered from a Supplier, the Supply Contract must be written and provide that the property in the materials shall pass unconditionally to the Contractor (or Sub-Contractor) by the time the items are ready for incorporation in the Works and have been properly stored for that purpose (Clause 30.3.2 and 30.3.3).

.5 where a Sub-Contractor is involved, the Sub-Contract must be in writing and provide, expressly, that the property in the materials passes immediately from the Sub-Contractor to the Contractor.

.6 where materials are manufactured or assembled by a Sub-Contractor:
(a) the Sub-Contract must be in writing, and
(b) the property in the materials expressly must pass to the Contractor as in Clause 30.3.4.

.7 the materials are in accordance with the Contract.

.8 the Contractor provides the Architect with reasonable proof that:
(a) the property in the materials is in him (the Contractor owns them) and
(b) the relevant contractual conditions have been met.

.9 the Contractor provides the Architect with reasonable proof that the materials are properly insured for their full value against 'Specified Perils' from the passing of the property in them to the Contractor, to their arrival on site (or adjacent thereto)

Note: Reasonable proof is normally the Architect's inspecting the Sub-Contracts and insurance policies and the physical materials.

See also Practice Note 5.

30.4 .1 The rules for Retention deduction by the Employer in any Interim Certificate are:

.1 the percentage is 5% or as otherwise specified in the Appendix. Where the Contract Sum is expected to be £500,000 or more (at Tender stage) the retention percentage should be no greater than 3%.

Thus, for most major projects, it is reasonable to anticipate that Retention will be 3%; so far, however, the implementation of a recommended lower Retention percentage has been rather restricted.

The advocating of a lower Retention percentage for major projects recognises the importance of cash flow in the industry but affords less protection for the Employer against such problems as insolvency of the Contractor. At Practical Completion (of the project or a part thereof) one half of the Retention held is released. In any Interim Certificate following Practical Completion only one half of the Retention Percentage may be deducted and held (Clause 30.4.1.3). Thus, if Retention is stated as 5%, once Practical Completion has been certified only 2½% Retention may be held, the first moiety (2½%) being released via the Interim Certificate following Practical Completion.

.2 Retention is deducted, as stated, against
(a) work which has not reached Practical Completion, and
(b) materials and goods (Clauses 20.2.1.2, 30.2.1.3, 30.2.1.4).

.3 Half the Retention Percentage may be deducted in respect of work
(a) which has reached Practical Completion, *but*
(b) has not achieved a Certificate of Completion of Making Good Defects, *or*
(c) an Interim Certificate to release Retention to a Nominated Sub-Contractor - Final Payment to the NS/C.

.2 'Contractor's Retention' - amounts deducted against Contractor's items.

'Nominated Sub-Contract Retention' - amounts deducted against Nominated Sub-Contractors' items.

30.5 Rules relating to Retention:

.1 The Employer's interest in any Retentions held is

'fiduciary as trustee'. There is no obligation upon the Employer to invest the monies so held.

In reality, the Employer probably will invest the funds but any interest earned thereby will accrue to him, not to the Contractor or Nominated-Sub-Contractor against whom the funds have been retained.

Re Tout & Finch Ltd. (1954) provides a precedent for this contractual provision.

.2.1 A statement of the Contractor's Retention and Nominated Sub-Contractors' Retention must be prepared by the Architect (or Q.S. under his instruction) on the date of each Interim Certificate.

.2 The statement must be issued to

(a) Employer
(b) Contractor
(c) all relevant Nominated Sub-Contractors

The statement will thus indicate how much Retention is to be deducted in arriving at the amount due. It will typically show the Retention deducted in total and against each Nominated Sub-Contractor.

.3 (Not in L.A. Edition). The Employer must:

(a) at the request of the Contractor or Nominated Sub-Contractor, place the Retention held against that party in a separate bank acount.
(b) certify to the Architect (copy to the Contractor) the amount so placed.

Note: Rayac Construction Ltd. v. Lampeter Meat Co. Ltd. (1979), established that retention held by an Employer is held in trust - it is *not* the Employer's money - and so should be set aside as a separate trust fund. This is vital should the Employer go into liquidation - the retention is *not* available to satisfy the Employer's creditors.

Note: Any interest on the sums so placed accrue to the Employer.

.4 If the Employer exercises his rights of set-off against Retention held (as specifed by Clause 30.1.1.2), he must inform the Contractor accordingly specifying the amount of set-off by reference to the latest issued statement of Retention.

This Clause is probably a result of the ruling

in the case of Gilbert Ash (Northern) Ltd. v. Modern Engineering (Bristol) Ltd. (1973), which reversed the decision of the case of Dawnays Ltd. & F. G. Minter Ltd. v. Trollope & Colls Ltd. (1971).

The Gilbert Ash decision related to a main Contractor - Sub-Contractor relationship in which the Sub-Contract contained certain special terms including those relating to payments and set-off. However, the principle laid down by the ruling is applicable to the J.C.T. Form. If one party has a claim against another giving rise to a situation involving contra-charges and set-off, the claim should be quantified as soon as possible, even if approximately, to permit valid set-off. The other party should immediately be given details of the claim and its quantification and afforded the opportunity of agreeing, challenging or somehow reaching a 'compromise' settlement.

Prior to final agreement and settlement of such a claim, any amount set-off should, it is recommended, be lodged with an independent stake-holder. This principle is applicable to main and Sub-Contract relationships.

See also Mottram Consultants Ltd. v. Bernard Sunley & Sons Ltd. (1975).

Retention must be released:
Half the relevant holding (whole or part of the Works dependent upon the scope of the Certificate of Practical Completion - whole or part) upon payment of the next Interim Certificate following the Practical Completion.

The remainder of the relevant holding of Retention (whole or part) upon payment of the next Interim Certificate following the latter of

(a) expiry of D.L.P., *or*
(b) issue of the Certificate of Completion of Making Good Defects.

30.6 .1.1 The Contractor must send all documents necessary to properly ascertain any adjustments to be made to the Contract Sum to the Architect (or Q.S. if so instructed by the Architect)

(a) before, or within a reasonable time of, Practical Completion

(b) including documents relating to the accounts of Nominated Sub-Contractors and Nominated Suppliers.

.2 The Quantity Surveyor must prepare a statement of all the final valuations of Variations (under Clause 13) if:

(a) the required information has been provided (see Clause 30.6.1.1), then

(b) within the Period of Final Measurement and Valuation as stated in the Appendix - 6 months from the date of Practical Completion, if no other period is specified.

The Architect must send a copy of this statement to:

(i) the Employer

(ii) the Contractor

(iii) a relevant extract to each Nominated Sub-Contractor.

.2 Method for adjustment of the Contract Sum:

Deduct:

.1 (a) all P.C. Sums in B.Q.

(b) all other sums in the B.Q. relating directly to NS/Cs as Clause 35.1

(c) all Contractor's profit in B.Q. on P.C. Sums and NS/Cs as otherwise included.

Note: In the L.A. Form the numbering is altered; an additional Clause is included as Clause 30.6.2.2: 'any amount due to the Employer under Clause 22A.2, 35.18.1.2 or 35.24.6'.

.2 (a) all provisional sums in B.Q.

(b) all items marked 'provisional' in B.Q.

Note: Under S.M.M.6 (b) should not arise except in a form of approximate quantities bill or section.

.3 (a) all valuations of omission Variations

(b) all valuations of work, the conditions of which have been changed due to Variations (Clauses 13.5.2, 13.2, 13.5.5 and 13.5).

.4 any Fluctuations amounts due to the Employer (Clauses 38, 39 and 40).

.5 'any other amount which is required by this Contract to be deducted from the Contract Sum.'
Thus Clause 30.6.2.5 may be regarded as a 'longstop' provision - to cover omissions of a more minor, less frequent nature. One's attention is drawn to the major areas of omission by the preceding 'check-listing' of the relevant areas.

.3 The Contractor must be given a copy of the Final Account before the issue of the Final Certificate.

30.7 The Architect must issue an Interim Certificate stating the amounts of all Nominated Sub-Contractors' Final Accounts. This must be done at least 28 days before the issue of the Final Certificate, even if this means that the one month interval between Interim Certificates is violated.

30.8 The Architect must issue a Final Certificate and inform each Nominated Sub-Contractor of its date of issue. The date of issue will be:

As soon as possible but before the expiry of 3 months from the latest of (in the L.A. Form reference is made to a period to be stated in the Appendix):

(a) end of the D.L.P. (as per the Appendix), or
(b) completion of making good of defects, or
(c) receipt by the Architect or Q.S. of the documents required to prepare the Final Accounts (Clause 30.6.1.1).

The Final Certificate must state:

.1 'the sum of the amounts already stated as due in Interim Certificates, *and*

.2 the Contract Sum adjusted as necessary in accordance with Clause 30.6.2.'

the difference between the two sums being expressed as a balance due (either way). From the 14th day after the date of the Final Certificate the balance is a debt due.

This is without prejudice to any other rights of the Parties.

30.9 .1 Subject to the provisions regarding legal action or Arbitration being instigated prior to the issue of the Final Certificate or within 14 days of its issue (and excepting instances involving fraud), the Final Certificate shall, in any proceedings due to the Contract, be conclusive evidence that:

.1 work, goods and materials, stipulated to be to the Architect's satisfaction, so comply, *and*

Additions:

.6 Nominated Sub-Contractors' final accounts in accordance with the NSC/4 or 4a.

.7 final account for the Contractor's work as a Nominated Sub-Contractor (Clause 35.2).

.8 final account for Nominated Suppliers' items (to include for 5% *cash* discount to the Contractor, as Clause 36) *excluding* any V.A.T. input tax to the Contractor. This acknowledges that some items may be subject to V.A.T. which a Contractor may not treat as recoverable input tax.

.9 Contractor's profit on Nominated or P.C.'d items at B.Q. rates (or as agreed, if appropriate, P.C. arising from a Provisional Sum Expenditure).

.10 amounts payable to the Contractor by the Employer due to

(a) Clause 6.2 - fees or charges
(b) Clause 7 - errors in levels and setting out.
(c) Clause 8.3 - inspection and testing
(d) Clause 9.2 - royalties
(e) Clause 17.2 - making good defects
(f) Clause 17.3 - making good defects
(g) Clause 21.2.3 - premiums paid by Contractor

.11 amount of the valuation of Variations of additions and of bill work where Variations have caused the conditions to change (Clause 13.5, 13.5.5).

.12 authorised expenditure against:

(a) Provisional Sums in the B.Q.
(b) Provisional Items in the B.Q. - under S.M.M.6 these should *not* arise except as noted above

.13 claims under:

(a) Clause 26.1 - direct loss/expense due to delays
(b) Clause 34.3 - direct loss/expense due to antiquities.

.14 (Not L.A. Editions) any amounts paid by the Contractor in respect of insurance against All Risks or 'Specified Perils'.which should have been effected by the Employer (Clause 22B or C).

.15 Fluctuations paid by the Contractor and recoverable from the Employer under the Contract (Clause 38 or 39 or 40).

.16 'any other amount which is required by this Contract to be added to the Contract Sum.'

.2 all computations in determining the amount of the Final Account are correct, except for any mistakes whether arithmetical or due to incorrect inclusion or exclusion of items.

This Clause is not really as meaningless as it appears to be at 'first glance'. Following the case of P. & M. Kaye Ltd. v. Hosier & Dickinson Ltd. (1972) and the wording of the contractual provisions, the Final Certificate is evidence:

(a) of items complying with the Contract where they are required to satisfy the Architect, *and*

(b) innocent and undiscovered mistakes in computations are accepted by the parties.

.2 If legal proceedings or arbitration have been instigated prior to the issue of the Final Certificate, its effect as conclusive evidence is modified:

.1 by the award of the judgement or settlement, *or, if earlier*

.2 by the elapsing of a period of 12 months during which neither party has taken any further action in the proceedings, then by any partial settlement reached.

.3 If arbitration or legal proceedings are commenced within 14 days of the issue of the Final Certificate, it shall *not* retain its function as conclusive evidence regarding any matter referred to the proceedings.

30.10 No certificate of the Architect, except the provisions relating to the Final Certificate, means that any items covered by that Certificate are in accordance with the Contract.

Thus, only the Final Certificate provides any evidence of items' compliance with the contractual requirements.

Note: Following Arenson v. Arenson (1977), if an architect negligently certifies too little, the contractor may sue the Architect.

Clause 31: Finance (No. 2) Act 1975—statutory tax deduction scheme

See also Practice Note 8.

This Act, the provisions of which affect all payments made on or after 6th April 1977, was introduced as the second major stage of measures to combat tax evasion, usually by labour-only Sub-Contractors, i.e. 'to get rid of the lump'.

It applies to all payments between a 'contractor' and 'sub-contractor', as defined by the Act.

31.1 Definitions:

'the Act' - Finance (No.2) Act 1975
'the Regulations' - Income Tax (Sub-Contractors in the Construction Industry) Regulations 1975, S.I. no. 1960
'Contractor' - a person who is a contractor for the purposes of the Act
'Evidence' - evidence required by the Regulations to be produced to a contractor to verify a sub-contractor's tax certificate
'Statutory Deducation' - a deduction under S69(4) of the Act or current amendments thereto
'Sub-contractor' - a person who is a sub-contractor for the purposes of the Act and the Regulations
'Tax Certificate' - a certificate issuable under S70 of the Act.

A 'contractor' is:

(a) any firm with a permanent building department, which

carries out 'construction operations'

(b) any local authority

(c) any development corporation or new town commission

(d) The Commission for New Towns

(e) The Housing Corporation, a housing corporation, etc.

31.2 .1 Clauses 31.3 to 31.9 do not apply if the Employer is stated *not* to be a 'contractor' in the Appendix. This, in effect, excludes the provisions of this Clause in connection with the Employer/Contractor relationship.

.2 If the Employer is initially not a 'contractor' but becomes one during the course of the Contract, the provisions of Clause 31 shall apply from that instant.

31.3 .1 Not later than 21 days before the first payment is due under the Contract (or before the first payment is due after the Employer has become a 'contractor'):

.1 the Contractor must inform the Employer, and provide the necessary evidence, that he is entitled to be paid without the Statutory Deduction, or the Contractor must

.2 inform the Employer in writing (copy to the Architect) that he may not be paid without the Statutory Deduction.

.2 If the Employer is not satisfied with the validity of the Contractor's evidence, he must notify the Contractor in writing within 14 days of the submission of the evidence of the intention to make the Statutory Deduction including a statement of the reasons for that decision. See also Clause 31.6.1 with which the Employer must comply.

31.4 .1 If a Contractor is not entitled to be paid without the Statutory Deduction but subsequently becomes so entitled, he must immediately inform the Employer.

i.e. If initially a Contractor does not have a Tax Certificate, but obtains one from the Inland Revenue during the course of the Contract.

.2 If the Contractor's Tax Certificate expires before the final payment is made, he must, prior to 28 days before the expiry date either

.1 provide valid evidence of non-deduction to the Employer

(see Clause 31.3.2 if the Employer is not satisfied therewith), or

.2 inform the Employer, in writing, that after the date of expiry, he is not entitled to be paid without the Statutory Deduction being made.

.3 The Contractor must immediately inform the Employer, in writing, if his current Tax Certificate is cancelled, stating the date of cancellation.

31.5 The Employer (as a 'contractor') must promptly send all vouchers given him by the Contractor (as a 'sub-contractor') to the Inland Revenue, in compliance with the Regulations.

31.6 .1 If, at any time, for any reasonable cause, the Employer believes he will have to make a Statutory Deduction from any payment due to the Contractor, he must immediately notify the Contractor in writing and require the Contractor to state the costs of materials (in total, i.e. including those of Sub-Contractors) to be included in any future payments.

The Contractor must furnish the information:

(a) not later than 7 days before each future payment becomes due, or
(b) within 10 days of the notifications, whichever is the later.

The Employer should check that:

(a) the items costed as materials are so permitted by the legislation - booklet IR 14/15(1980), and
(b) the cost of the materials is reasonable

.2 Upon the Contractor's giving such notice he must indemnify the Employer against any direct loss/expense caused by his incorrect statements.

.3 If the Contractor does not give such notice, the Employer may himself make a fair estimate of the direct costs of materials involved. (In practice, the Q.S. will probably do this for the Employer.)

It is essential for the Employer to secure such indemnity (Clause 31.6.2) from the Contractor as any Statutory Deductions incorrectly *not* made by the Employer will, under the Act, be his liability to pay to the Inland Revenue.

31.7 If any error or omission in the Statutory Deductions is made, the Employer must, under the Statutory Obligations, effect the correction by a payment to or from the Contractor.

31.8 Clause 31 to prevail over any other conditions with which the operation of this provision causes conflict, i.e. the provisions of this Clause take precedence over other terms of the Contract, if in conflict with them.

31.9 Arbitration must be used, as Article 5, to resolve any dispute under this Clause between the Employer, the Architect on his behalf and the Contractor. The only exception is where there is statutory provision dictating another method of resolution.

Evidence is normally provided by submission of a Tax Certificate.

Three types of Tax Certificate are currently in use in the industry - 714I, 714P, 714C (issued by Inland Revenue)

714I - Individual Certificate
Contains an authenticated photograph and signature of the holder, is numbered and states the date of expiry. It must itself be presented as evidence, for verification.

714P - Partnership Certificate (also for small firms)
Contains an authenticated photograph of the director or secretary, the signature of that person, the name of the firm for which that person is acting, is numbered and states the date of expiry. It must itself be presented as evidence for verification.

In examination of both the I and P certificates, proof of identity of the presenter should be sought by signature and/or other evidence.

714C - Company Certificate (larger and and limited companies)
No photograph. Operates without the use of vouchers. It has the company's registration number stamped on it. It bears the signature of the company secretary. It contains the name of the company, is numbered and states the date of expiry.

For a C certificate an official copy may validly be used for verification purposes.

Note: As this legislation is intended to stamp out 'lump labour', the materials components of payments are not subjected to deductions.

Clause 32: Outbreak of hostilities

A footnote to the Contract indicates that, should circumstances so dictate, the parties may agree alternative measures and procedures to those expressed in this Clause.

32.1 It is not necessary for war to be declared involving the U.K. but there must be:

(a) general mobilisation of U.K. armed forces, occurring
(b) during the currency of the Contract to allow either party to determine the employment of the Contractor by (written) notice, by registered post or recorded delivery.

The notice may be given only:

.1 after 28 days from the date of general mobilisation, *or*
.2 before Practical Completion of the Works, *or* after Practical Completion if the Works have sustained war damage under Clause 33.4.

32.2 The Architect may issue A.I.s

(a) requiring the Contractor to execute protective work as specified, *and/or*

(b) continue the Works to a specified state of completion provided the A.I.s are given within 14 days of the Employer's giving or receiving the determination notice, then they are valid.

In such an instance, the Contractor must execute the A.I. (as if notice of determination had not been given).

If the situation renders the completion of the A.I. work impracticable (for reasons outside the control of the Contractor) within 3 months of their being given, the Contractor may abandon the work involved.

32.3 After

(a) 14 days from the Employer's giving or receiving a determination notice, *or*
(b) the execution of A.I.s under Clause 32.2, *or*
(c) the abandonment of Clause 32.2 A.I. work, the Contractor shall be paid for the work executed:

(i) Contract work - under Clause 28.2 (not Clause 28.2.2.6)

(ii) Clause 32.2 A.I. work - under the Variations *and* provisions (Clause 13.2, 13.5).

Clause 33: War damage

Clause 33.4 defines war damage by reference to S2 of the War Damage Act 1943 or its current equivalent.

33.1 If war damage occurs to the Works or any 'materials or goods on site':

.1 such damage is ignored in the computation of amounts payable to the Contractor

.2 the Architect may issue A.I.s regarding

(a) removal/disposal of debris
(b) removal/disposal of damaged work
(c) execution of protective work

.3 the Contractor to make good war damage, proceed with and complete the Works. The Architect to give a fair and reasonable Extension of Time

.4 work involved in A.I.s under Clause 33.1.2 to be valued as Variations under Clause 13.2 and 13.5.

33.2 If notice of determination is given after war damage has occurred:

(a) 'protective work' in Clause 32 is to include the scope of A.I.s under Clause 33.1.2
(b) A.I.s issued under Clause 33.1.2 and given prior to the Employer's giving or receiving a notice of determination

and uncompleted are then deemed to be A.I.s under Clause 32.2.

The effect of (b) is that, if possible, the work required by the A.I.s must be completed despite determination. This is obviously reasonable on grounds of safety.

33.3 The Employer is entitled to claim from any Governmental war damage fund for compensation in respect of war damage to his property in the Works, i.e. the Works as completed or partly completed and any items on or off site for which he had paid (and the property in them is in the Employer).

33.4 Defines war damage, as above. (S2 of the War Damage Act, 1943).

Clause 34: Antiquities

34.1 All fossils and antiquities discovered on the site during the execution of the Works are the property of the Employer.

The Contractor, upon discovering a fossil or antiquity, must immediately:

.1 use his best endeavours not to disturb the object. He must stop any work likely to disturb or harm the object or impede its removal or inspection. This may obviously delay the project and thereby give grounds for an Extension of Time (Clause 25.4.5.1) and direct loss/expense (but see Clause 34.3.1)

.2 take steps as thought necessary to preserve the object in the exact location and condition as it was discovered

.3 inform the Architect or Clerk of Work of the discovery and its exact location.

34.2 The Architect must issue an A.I. about action concerning the object discovered. Here the Architect is empowered to require the Contractor to allow others onto the site to examine/remove the object. Such a party (probably an archaeologist) is for insurance purposes (Clause 20) considered to be an 'Artist or Tradesman'.

34.3 .1 The Contractor's direct loss/expense due to antiquities is reimbursable at the exercise of the Architect's opinion - the Architect, or Q.S. at his instruction to determine the amount of direct loss/expense.

.2 If necessary for the calculation of the direct loss/expense incurred, the Architect must state, in writing to the Contractor, any period of Extension of Time awarded due to the discovery of antiquities.

.3 Any amounts of direct loss/expense are added to the Contract Sum (see above for details).

CONDITIONS: PART 2: NOMINATED SUB-CONTRACTORS AND NOMINATED SUPPLIERS

Clause 35: Nominated Sub-Contractors

See also Practice Notes 10, 11, 12 and 13.

In many ways this part of the Conditions of Contract represents also a code of practice. It is thus rather more lengthy and complete than any preceding documents regarding this somewhat complex subject.

35.1 Defines a Nominated Sub-Contractor as one:

where 'the Architect has, whether by the use of a prime cost sum or by naming a sub-contractor, reserved to himself the final selection and approval of the sub-contractor...'

The reservation and thence nomination may occur in:

.1 The Contract Bills

.2 An A.I. (under Clause 13.3) regarding the expenditure of a provisional sum in the B.Q.

.3 An A.I. (under Clause 13.2) requiring a Variation where:

  .1 it comprises additional work to that in the Contract Drawings or Bills, *and*

  .2 Nominated Sub-Contract items of such additional work as are of a similar type to that indicated to be subject to Nomination of a sub-contractor in the Contract Bills.

(Similar may be read here as, it is suggested, almost identical. This Clause is of obvious importance where the original Sub-Contract work is complete, the additional work constituting, in effect, a new Sub-Contract.)

This provision is a formal recognition of the usual practice in industry. It does represent a modification of the general principle that only where a P.C. Sum exists in the B.Q. or arises due to A.I.s regarding the expenditure of a provisional sum in the B.Q. may a Nomination occur. This Clause should make the situation clear, even to the more pedantic.

This represents recognition that it is reasonable for the scope of NS/Cs' work to be extended within their specialist areas if so required by the Variations to the project.

.4 An agreement (not unreasonably withheld) between the Contractor and the Architect on the Employer's behalf. This could include a nomination against the Contractor's measured items in the B.Q.

Any Sub-Contractor selected by this procedure must be nominated in accordance with the provisions of this Clause and is termed a Nominated Sub-Contractor.

This Clause applies notwithstanding the provisions of S.M.M.6, Clause B9.1, which dictates that there must be a P.C. Sum in the B.Q. for a Nominated Sub-Contractor to be used.

It is usual for a B.Q. to contain, in relation to each NS/C:

(a) P.C. Sum (inserted by the Q.S.)
(b) Contractor's Profit thereon (priced by the Contractor)
(c) General Attendances on NS/C (priced by the Contractor)
(d) Other Attendances on NS/C (details by the Q.S., priced by the Contractor)
(e) Builder's Work in Connection - measured items to be executed, and priced, by the Contractor.

35.2 Contractor tendering for nominated work.

.1 For this situation to arise the following must apply to the work in question:

(a) the Contractor carries out such work directly, in the usual course of his business
(b) the work is included in the Contract Bills and Clause 35 is applicable (see Clause 35.1)
(c) the items of work are set out in the Appendix (by the Contractor)
(d) the Architect is prepared to receive the Contractor's tenders for such items.

The Employer retains the right to reject any tender.

The Contractor may not sub-let any work so obtained without the consent of the Architect.

For the purposes of this Clause, any item against which the Architect intends to nominate an NS/C, arising through A.I.s regarding expenditure of provisional sums in the B.Q. (Clause 13.3), is deemed to have been set out in the B.Q. and included in the Appendix.

The effect is merely that the Contractor may, if the Architect allows, submit a tender for specialist work which he ordinarily executes, whenever such work arises from an A.I. expending a provisional sum and the Architect wishes to nominate.

.2 Relates to some formal documentation provisions for the purposes of a Contractor's accepted tender in respect of Variations - to refer to Clause 13 wherein references to Contract Drawings and Contract Bills are deemed to refer to the relevant Tender Documents under Clause 35.2.

.3 Only the provisions of Clause 35.2, not all of Clause 35, are applicable to a Contractor whose tender is accepted under this Clause.

35.3 Lists and identifies documents relating to NS/Cs under the Standard Form of Contract. The usual set will be:

NSC/1	Tender	The 'basic' method.
NSC/2	Employer - NS/C Agreement	
NSC/3	Nomination	
NSC/4	Sub-Contract	

If NSC/1 is not used, NSC/1, 2a and 4a will be the standard documentation - the 'alternative' method.

Procedure for Nomination of a Sub-Contractor:

35.4 Contractor's right to reasonable objection to a proposed NS/C.

.1 The over-riding provision that 'No person against whom the Contractor makes a reasonable objection shall be a Nominated Sub-Contractor'.

.2 Where NSC/1 and 2 are used:

Any Contractor's objection must be made
(a) as soon as possible, *but*
(b) not later than when he sends NSC/1 to the Architect (Clause 35.10.1)

.3 Where NSC/1 and 2 are *not* used; (Clause 35.11 and 35.12 apply).

Any Contractor's objection must be made:
(a) as soon as possible, *but*
(b) not later than 7 days from his receipt of the nominating A.I. (Clause 35.11).

35.5 .1.1 NSC/1 and 2 must be used in connection with nominations unless:

.2 the B.Q. or A.I. under
Clause 13.2 - Variations
Clause 13.3 - Expenditure of a provisional sum
Clause 35.3.2 - Contractor tendering
states that NSC/1 and 2 will not be used (Clauses 35.11 and 35.12 to apply) for that nomination.
The Document (B.Q. or A.I.) must then state if NSC/2a is to be used.

.2 Regarding any work to be nominated, the Architect may issue an A.I.
(a) substituting Clause 35.11 and 35.12 for NSC/1 and 2
or
(b) substituting NSC/1 and 2 for Clauses 35.11 and 35.12.

Any such A.I.
(i) is to be treated as a Variation (Clause 13.2), *and*
(ii) may not be given after the preliminary notice of nomination (Clause 35.7.1) or a nomination

A.I. (Clause 35.11) has been issued *except* for the purposes of re-nomination (Clause 35.23 and 35.24).

Use of NSC/1 and NSC2:

35.6 Only persons who have tendered on NSC/1 and entered into Agreement NSC/2 may be nominated unless Clauses 35.11 and 35.12 apply.

35.7 .1 Where NSC/1 and 2 are used, the Architect must send a copy of these completed documents to the Contractor with a preliminary notice of nomination.

The Contractor must 'forthwith' settle with the proposed Sub-Contractor any items which remain to be agreed as set out in the Particular Conditions of Schedule 2 of NSC/1. This must be done before the Architect can nominate under Clause 35.10.2.

.2 Upon receipt of the documentation, the Contractor must settle the outstanding items of the Particular Conditions (as Clause 35.7.1).

35.8 If within 10 working days from receipt of the preliminary notice of nomination the Contractor has been unable to settle all the Particular Conditions with the proposed Sub-Contractor, he must:

(a) continue to try to settle them (Clause 35.7.2), *and*
(b) inform the Architect, in writing, of the reasons for the inability to settle.

Then the Architect must issue A.I.s as he considers necessary.

35.9 If a proposed Sub-Contractor informs the Contractor that he is withdrawing his offer made under NSC/1, the Contractor must immediately inform the Architect, in writing, and do nothing further until he receives appropriate A.I.s.

35.10 .1 As soon as settlement is reached between the proposed Sub-Contractor and the Contractor, the Contractor must send the duly completed NSC/1, in total, to the Architect.

.2 When the Architect receives the completed NSC/1, he

must issue the nomination A.I. to the Contractor on NSC/3, with a copy to the Sub-Contractor.

NSC/1 and NSC/2 not used

35.11 Clause 35.5.1.2 applies

.1 the Employer must complete NSC/2a with the proposed Sub-Contractor (unless agreed and tendered to the contrary),

.2 the Architect must issue a nominating A.I. to the Contractor, copy to the NS/C.

35.12 Within 14 days of the nomination A.I. (Clause 35.11), the Contractor must execute Sub-Contract NSC/4a with the NS/C.

Payment of a Nominated Sub-Contractor:

35.13 .1 Upon the issue of each Interim Certificate, the Architect must:

.1 direct the Contractor as to the amounts of any (Interim or Final) Payments to NS/Cs included, as computed by the Architect in accordance with the relevant provisions of NSC/4 or 4a. This will usually be in the form of a schedule attached to the Contractor's copy of the Interim Certificate.

.2 inform each NS/C of any payments so directed (Clause 35.13.1.1).

.2 The Contractor must make the payments to the NS/Cs in accordance with the relevant NSC/4 or 4a provisions. (NSC/4 - Clause 21.3.1.1 - within 17 days of the date of issue of the Interim Certificate, less 2½% cash discount - applies also to NSC/4a.)

.3 Prior to the issue of the second and subsequent Interim Certificates, and the Final Certificate, the Contractor must provide the Architect with reasonable proof that payment to any NS/Cs (as required by Clause 35.13.2) has been discharged, usually by production of a receipt detailing the amounts involved. Set-off may form all/part of the due discharge.

.4 If the NS/C fails to give a receipt or other document to enable the Contractor to prove payment and the Architect is satisfied that the Contractor has paid, the provisions regarding proof of payment (Clause 35.13.3) are deemed to be satisfied (Clause 35.13.5 does not apply); in this case the failure to prove must be the fault of the NS/C in failing to give evidence of payment.

.5.1 Where NSC/2 or 2a applies, the Employer must (if necessary) operate Clause 35.13.5.3 and 35.13.5.4, otherwise their operation is optional upon the Employer.

.2 If the Contractor fails to provide reasonable proof of payment to an NS/C (under Clause 35.13.3), the Architect must:

(a) issue a certificate to that effect, stating the amount involved, *and*
(b) issue a copy of that certificate to the appropriate NS/C.

.3 The Employer must pay the amounts due direct to the NS/Cs, in which case:

(a) future payments to the Contractor must be reduced by payments he has failed to make to NS/Cs.
(b) V.A.T. due to the NS/Cs in connection with such amounts must also be paid direct
(c) the rights of set-off are unaffected
(d) no such payments must be made if the Employer cannot set them off against amounts due from him to the Contractor.

This is always provided that
(i) the Architect has issued a non-payment certificate,
and
(ii) the proof failure is not due to the NS/C under Clause 35.13.4

Thus, in summary, direct payment must be implemented where:

- a certificate of non-payment has been implemented
- the proof failure is due to the Contractor's default
- amounts due to the Contractor more than cover any amounts due to NS/Cs after any exercise of set-off rights between the Employer and Contractor for other issues
- NSC/2 or 2a applies.

.4 Direct payments provisions are further modified as follows:

.1 the set-off of the direct NS/C payment by the

Employer against amounts due to the Contractor under an Interim Certificate must occur:

(a) at the time of payment to the Contractor, if the set-off leaves any balance to be paid, *or*

(b) within the 14 day period of honouring (See Clause 30) if the set-off leaves no balance for payment to the Contractor.

.2 where the sum due to the Contractor comprises only Retention to be released, the limit of set off is the amount of Retention to be released to the Contractor in that payment.

.3 where the Employer is to pay 2 or more NS/Cs direct and the amount available, due to the set-off restrictions, is insufficient to meet the payments in full, the Employer must *pro rata* the amount available (or otherwise fairly apportion the available sum). Any further future amounts must be treated in a similar manner by the Employer to pay NS/Cs direct (e.g. 1st and 2nd releases of Retention).

.4 if the Contractor's business is being wound up by either:

(a) a Petition presented to the Court, *or*

(b) a winding up resolution having been passed (unless for amalgamation or reconstruction)

the set-off provisions against the Contractor shall not apply; the relevant time in such instances is if either the first of (a) or (b) has occurred at the time when (under Clause 35.13.5.3) the Employer is to execute the reduction and direct payment to the NS/C.

<u>Note</u>: This Clause will require amendment if the Contractor is not a person to whom the law of insolvency of a company is applicable. In such cases, bankruptcy law applies.

35.14 Extension of Time to Nominated Sub-Contractors

.1 The Contractor's authority to grant Extensions of Time to NS/C s is limited to the Extensions being:

(a) in accordance with the provisions of NSC/4 or 4a, *and*

(b) with the written consent of the Architect.

This applies also to projects to be completed in parts by the NS/Cs.

.2 Clause 11.2.2 of NSC/4 or 4a is applicable. The NS/C and the Contractor must send written particulars and estimate of the period involved to the Architect requesting an Extension of Time for the NS/C. The Architect must give his written consent before the Contractor can award any Extension of Time to the NS/C.

35.15 Failure to Complete Nominated Sub-Contract Works

.1 If any NS/C fails to complete the Sub-Contract Works (or part thereof, if applicable) within the relevant time (Sub-Contract period plus any Extensions), the Contractor should so notify the Architect (copy to NS/C).

Provided the Architect is satisfied that the Extension of Time provisions have been met (Clause 35.14), he must certify to the Contractor (duplicate to NS/C) that the NS/C has failed to complete by the appropriate date.

.2 The certificate must be issued within 2 months from the date of notification of the NS/C's failure to complete.

35.16 Practical Completion of Nominated Sub-Contract Works

The Architect must issue a Certificate of Practical Completion of the NS/C Works when such has been achieved *in his opinion* (duplicate to NS/C). Only one such Certificate is required even if the NS/C Works are completed in sections.

The practical completion is deemed to have occurred on the date of the certificate for the purposes of:
Clause 35.16 - Practical Completion
Clause 35.17, .18, .19 - final payment
Clause 18 - partial possession by Employer.

<u>Westminster City Council v. J. Jarvis & Sons Ltd. (1970)</u> (House of Lords) - Jarvis used Peter Lind to execute piling as a Nominated Sub-Contractor who purported to complete their work on the due date. One month later an excavator of Jarvis accidentally knocked a pile which broke off. Tests revealed many piles to be defective due to bad workmanship and/or materials.

In the ensuing claims and legal action over the cost of

remedial works, liquidated damages liability and delays, the following was decided:

(a) Lind had achieved Practical Completion of their work on the due date (when they also withdrew from the site).
(b) The piles had latent defects and so the Sub-Contractors were in breach, not delay.
(c) Lind returned to remedy a breach, not to fulfil (i.e. complete) their Contract.
(d) No Extension of Time could be awarded and the Employer therefore had a valid claim against the Contractor (Jarvis) for liquidated damages for delay.

Lind were obviously resolved to bear the liability which, the tests revealed, was clearly theirs. They would have to pay either the Employer or the Contractor and so allowed those parties to resolve the legal issue, having executed the necessary remedial work.

As the case was decided, the Contractor was the party to seek redress against them.

It is, perhaps, worth considering an instance where the defects had been discovered prior to Practical Completion of the Sub-Contract Works. Here it is probable that Practical Completion would have been delayed thereby giving the Contractor grounds for an Extension of Time for the main Contract Works. The Employer could then not claim liquidated damages against the Contractor.

J. Jarvis Ltd. v. Rockdale Housing Association (1985): piling was executed by a NS/C; tests revealed defective piles. Held that a technical breach of the contract resulting from a substantial breach of sub-contract by a NS/C is not 'negligence or default of the Contractor'. (See NSC/4 Clause 13.2.)

35.17 Final Payment of Nominated Sub-Contractors

The Architect may secure final payment to an NS/C (subject to provisions of NSC/2 or 2a - Clause 5 or 4 - being unamended) by including the NS/C's final account in an Interim Certificate issued after Practical Completion of the NS/C Works. This is mandatory if 12 months has elapsed from the date of NS/C's Practical Completion, as certified.

The final payment is subject to provisos:

.1 the NS/C has made good the defects as required contractually, *and*
.2 has sent to the Architect or Q.S. (via the Contractor)

all the documents required for the preparation of the final account.

35.18 Upon payment by the Contractor to the NS/C of the certified final payment (as Clause 35.17):

.1.1 If the NS/C fails to make good any defect as required, the Architect must issue an A.I. nominating another Sub-Contractor to execute that work.

.2 The Employer must, as far as possible, under NSC/2 or 2a recover the sum due to the substituted NS/C from the original NS/C. If this cannot be done the Contractor must pay the appropriate sum to the Employer provided the Contractor had agreed to the substituted NS/C's price prior to that nomination.

.2 Clause 35.21 (NSC/2 or 2a and the Contractor) is to prevail over the provisions of this Clause.

The most likely application of Clause 35.18.1.2 is where an NS/C becomes insolvent after completing the Contract work and procuring final payment. Thus, wherever possible it should be ensured that all making good of defects has been satisfactorily completed prior to final payment being made.

35.19 Despite any final payment to an NS/C:

.1 The Contractor retains full responsibility for loss/damage in respect of items against which final payment to an NS/C has been made to the same extent as for items of uncompleted work, goods and materials up to the date of:

(a) Practical Completion of the Works, *or*, *if earlier*,
(b) the Employer's taking possession.

.2 The insurance provisions under Clause 22 remain in full force and effect.

35.20 Employer and Nominated Sub-Contractor.

(a) Nothing in the Conditions renders the Employer in any way liable to an NS/C.

(b) The only contractual relationship between the Employer and an NS/C is as specified in either NSC/2 or 2a, as applicable.

35.21 The Contractor and Clause 2 of NSC/2 or Clause 1 of NSC/2a (design, materials, performance specification).

The Contractor is 'not liable' to the Employer in any way regarding NS/C works governed by Clause 2 of NSC/2 or Clause 1 of NSC/2a. The Contractor does, however, retain his full contractual responsibilities regarding the supply of goods, materials and workmanship, including those of an NS/C.

Thus the Contractor is expressly excluded from any design responsibility which the NS/C may undertake. He does retain responsibilities for the work execution - i.e. that it complies with the design.

Note: All design is deemed to come to the Contractor from the Employer via his agent, the Architect. The Contractor undertakes no design responsibility.

35.22 Limitation of Liability of NS/C, or:

Any limitation of liability by the NS/C to the Contractor under Clause 2.3 of NSC/4 or 4a is to be passed on to the Employer.

Note: This may not be possible due to the protection afforded to consumer sales over those afforded to commercial sales by the Unfair Contract Terms Act, 1977 - but see under Gloucestershire County Council v. Richardson (1968).

35.23 Where Nomination does not proceed further, the Architect must either

(a) re-nominate, *or*
(b) issue a Variation (Clause 13.2) omitting the nominated work.

This would occur where: (.1 & .2 - 'basic' method; .1 & .3 - 'alternative' method)

.1 the Contractor sustains a reasonable objection to the proposed NS/C (Clause 13.4), *or*

.2 the NS/C does not settle the outstanding Particular Conditions of Schedule 2 of NSC/1 within a reasonable time when requested so to do by the Contractor (Clauses 35.6 to 35.10 and 35.7.1), *or*

.3 under Clauses 35.11 and 35.12, the NS/C fails to enter into NSC/4a within a reasonable time and without good cause.

35.24 Where Re-Nomination is necessary:

.1 if the Architect is of the opinion that the NS/C has made default regarding items as NSC/4 or 4a, Clauses 29.1.1 to 29.1.4 (grounds for determination) and the Contractor has informed the Architect of the alleged default(s) together with any observations of the NS/C in relation thereto, *or*

.2 the NS/C goes into liquidation, etc. (except for the purposes of reconstruction or amalgamation) *or*

.3 the NS/C determines his employment under Clause 30 of NSC/4 or 4a,

then the following are applicable.

.4 If Clause 35.24.1 is invoked,

.1 The Architect must issue an A.I. to the Contractor to give the NS/C notice specifying the default. That A.I. may also instruct the Contractor to obtain a further A.I. prior to determining the employment of the NS/C.

also

.2 Following the giving of the specifed notice, the Contractor must inform the Architect if the employment of the NS/C has been determined. If the second A.I. under Clause 35.24.4.1 has been given, the Contractor must confirm the determination to the Architect.

.3 Upon the Architect being finally informed of the determination (Clause 35.24.4.2) he must re-nominate as necessary. If such determination has been occasioned by the failure of the NS/C to make good defects (see also NSC/4 or 4a, Clause 29.1.3 and main Contract Clause 35.18), the Contractor must agree the price to be charged by the substituted NS/C (due to set-off provisions, as above).

.5 If Clause 35.24.2 is invoked,

The Architect must re-nominate as necessary. Provided that, if the receiver is appointed and the Architect reasonably believes that he is prepared to continue to completion of the relevant sub-Contract without in any way detrimentally affecting any other person involved with the Works, the Architect may postpone a re-nomination.

.6 If Clause 35.24.3 is invoked,
The Architect must re-nominate as necessary. Any excess price of the substituted NS/C over that of the original NS/C may be recovered by the Employer from the Contractor (by set-off, as a debt).

The responsibility of the Contractor to pay any excess price is reasonable as the determination of the original NS/C would be due to the Contractor's default.

.7 Any amounts payable to the re-nominated NS/Cs must be included in Interim Certificates and added to the Contract Sum.

Re-nomination under Clause 35.24.6 is the exception (obviously as above).

Bickerton v. N. W. Regional Hospital Board (1969): If the original NS/C 'drops out', the Employer must require the Architect to re-nominate. Subject to the provisions regarding execution of specialist work to be the subject of nomination by the Contractor's submitting a tender which is accepted, all items indicated to be executed by an NS/C must, in fact, be executed by an NS/C. The Contractor is not obliged to execute such items.

The costs of the original and subsequent Nominated Sub-Contract items fall on the Employer except any additional costs under Clause 35.24.6, which fall upon the Contractor.

Following Fairclough Building Ltd. v. Rhuddlan Borough Council (1985): An Architect's Instruction for re-nomination must include outstanding and remedial items of work; the main contractor has no duty or right to complete the work of the sub-contractor which failed.

The main contractor may reject a re-nomination instruction if the substituted sub-contractor will not undertake to complete the items of work in compliance with the main contractor's programme.

The Architect must re-nominate within a reasonable time of the main contractor's application for an instruction; if not, the delay in re-nominating will be a Relevant Event under Clause 25.4 and may give grounds for a loss and expense claim under Clause 26.

35.25 A.I.s *re* NS/C determination

The Contractor must not determine any Nominated Sub-Contract without an A.I. to do so.

35.26 If the determination is made under Clause 29 of NSC/4 or 4a, the Architect must direct the Contractor as to any amounts due to the NS/C in an Interim Certificate under Clause 29.4 of NSC/4 or 4a (balance paid to NS/C after set-off by the Employer and the Contractor and subject to usual 2½% cash discount).

Contractual Relationships

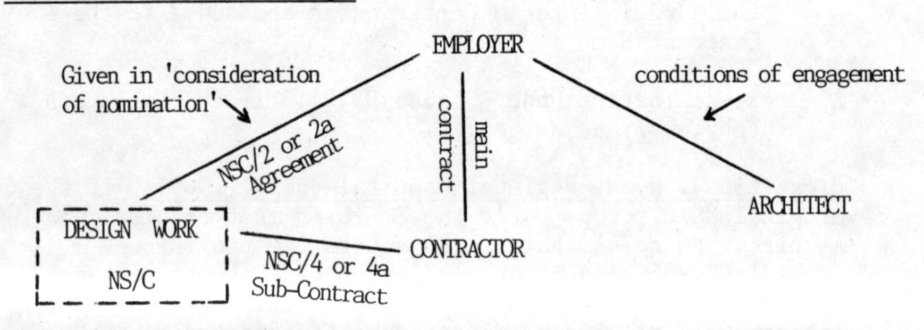

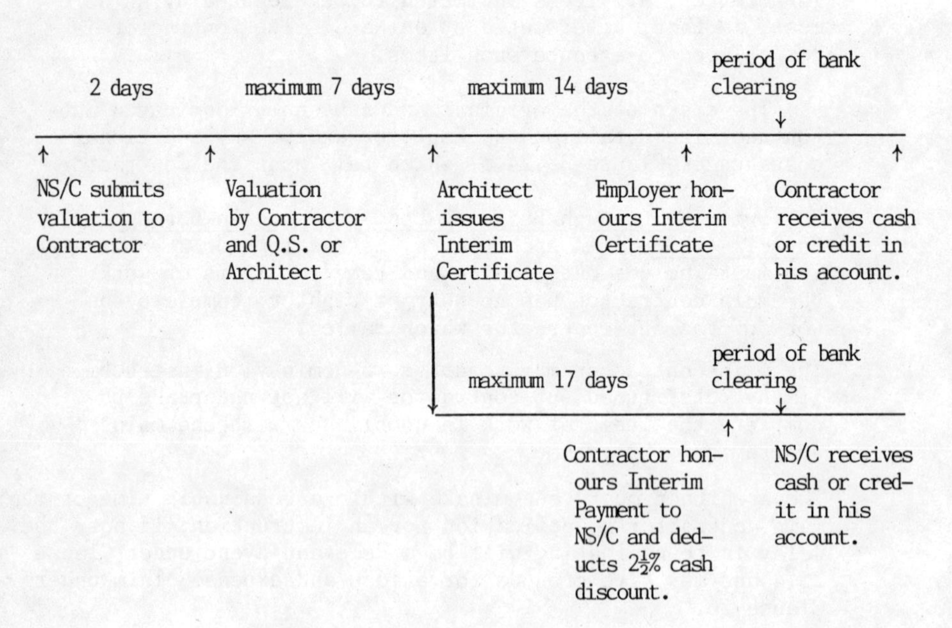

Additional noteworthy points regarding NS/Cs.

(a) If the Contractor determines the employment of an NS/C without the requisite A.I., he is in breach of Contract.

(b) If the Architect does not re-nominate as required, the Employer and Contractor should make a separate agreement regarding what is to be done, e.g. use of Artists and Tradesmen; the Contractor to execute the work. Payment will be by an agreed price or *quantum meruit*.

(c) If an NS/C repudiates the Sub-Contract the Contractor may seek to recover from that NS/C only damages he has incurred directly resulting from the repudiation.

(d) Any items to be executed by an NS/C are still the responsibility of the Contractor. He must ensure that they are carried out properly by the nominee (or original and subsequent nominees). If the Contractor fails to do this he is in breach of his Contract with the Employer.

Remember:

A Nominated Sub-Contractor is still a Sub-Contractor of the Contractor.
An NS/C is merely nominated by the Architect.
The Contractor is responsible for all work execution.
The Architect is responsible for design (from the Contractor's viewpoint).
The Contractor is entitled to $2\frac{1}{2}$% (1/39th of nett accounts to add) cash discount (no other) for prompt payment.

Clause 36: Nominated Suppliers

See also Practice Note 15.

36.1 .1 Definition of an NSup given.
NSup - nominated by the Architect to supply materials or goods, to be fixed by the Contractor.

Thus, it is usual for the B.Q. to contain in relation to each NSup:

(a) P.C. Sum (inserted by Q.S.)
(b) Contractor's Profit thereon (priced by Contractor)
(c) Fix items to be priced by the Contractor.

The nomination by the Architect may occur where:

.1 B.Q. contains an appropriate P.C. Sum and the name of the NSup, *or*
B.Q. contains an appropriate P.C. Sum and the Architect issues an A.I. of nomination subsequently (Clause 36.1).

.2 B.Q. contains a provisional sum. The A.I. regarding its expenditure gives rise to a P.C. Sum against which a supplier is nominated in that or a subsequent A.I.

.3 B.Q. contains a provisional sum. The A.I. regarding its expenditure effectively gives rise to a P.C. Sum - nominates the supplier if items to be purchased by the Contractor under that A.I. are available from one source of supply and only one supplier.

.4 Where a Variation (Clause 13.2) requires goods available from a sole source of supply and only one supplier, this is deemed to be a nomination of that supplier against a P.C. Sum.

Thus the P.C. Sum against which nomination takes place occurs:

(a) in the B.Q.
(b) by an A.I. *re* expenditure of a provisional sum
(c) as (b) but where there is only one Supplier, not necessarily named as a nomination
(d) by Variation where there is only one Supplier.

.2 Unless there is a sole Supplier of materials or goods, the supply of the materials or goods must be the subject of a P.C. Sum in the B.Q. for the Supplier to be nominated.

This recognises that it has been possible effectively to nominate a Supplier without formally doing so. This is no longer the case. If there is more than one Supplier and more than one source of supply (only one of each now constitutes a nomination), nomination may occur only by an A.I. regarding the expenditure of a P.C. Sum (either itself in the B.Q. or arising out of A.I.s regarding the expenditure of provisional sums contained in the B.Q.).

This is clearly demonstrated by:

36.2 The Architect must issue an A.I. nominating any supplier against any P.C. Sum (in the B.Q. or howsoever arising).

36.3 .1 To calculate amounts due in respect of NSup items (Clause 30.6.2.8) only 5% *cash* discount for prompt payment is allowed to the Contractor (1/19th to add to nett accounts), all other discounts, etc., must be passed on to the Employer.

The Employer must pay the gross amount due (including the cash discount allowance) to the Contractor, which also includes

.1 any tax or duty, charged under an Act of Parliament, on the goods supplied, not recoverable elsewhere under the Contract (exclusive of V.A.T. input tax which the Contractor may reclaim from the Customs and Excise)

.2 the nett cost of appropriate packaging, carriage and delivery; subject to any credits for the return of the packaging to the Supplier

.3 fluctuations, subject to the discount provisions as above.

.2 If the Architect believes that the Contractor has incurred (properly) any expense in obtaining the goods etc., which he would not be reimbursed under any other Contract provision, such expense must be added to the Contract Sum.

36.4 The Supplier must enter (or be prepared to enter) into an agreement with the Contractor containing the provisions set out below in order that he may be nominated by the Architect. (The only exception is where the Architect and Contractor agree otherwise.)

.1 The quality and standards of items supplied must be as specified or, if appropriate, to the reasonable satisfaction of the Architect.

.2 The NSup must make good or replace (at his own cost) items in which defects appear prior to the expiry of the D.L.P. The NSup must also bear any direct costs of the Contractor in consequence. This is qualified:

.1 where the items are fixed, a reasonable examination prior to fixing by the Contractor would not have revealed the defects

.2 the defects are due solely to defective workmanship or materials in the items supplied. It is invalidated if the defects were caused by someone outside the control of the NSup (e.g. inadequate storage by the Contractor).

.3 Delivery must be in accordance with a programme agreed between the Contractor and NSup, *or* in accordance with the Contractor's reasonable directions.

.4 The NSup must allow the Contractor 5% cash discount for payments made for items within 30 days from the end of the month in which those items were delivered (as discussed earlier, any other discounts must be passed on to the Employer).

.5 The NSup need not deliver any items after a determination of the Contractor's employment *except* any for which he has been properly paid.

.6 The operation of Clause 36.4.4 fully discharges the payment provisions.

.7 Ownership of the items passes to the Contractor upon their delivery, even if they have not been paid for. (However see Aluminium Industrie Vaasen BV v. Romalpa Aluminium Ltd. (1976) where a retention of title clause applied.)

.8 If disputes arise between the Contractor and NSup which are similar to a dispute between the Contractor and the Employer which has itself been referred to Arbitration, then the Contractor/NSup dispute must be referred to the same Arbitrator, under Article 5, whose decision shall be binding.

.9 Any Contract of sale conditions may not prevail over the conditions set out in this Clause.

Note: Use of JCT Standard Form of Tender by Nominated Supplies (TNS/1) is advisable.

36.5 .1 If any liability restriction or exclusion (as Clauses 36.5.2 or 36.5.3) exists between the NSup and Contractor and has the written approval of the Architect, the Contractor may restrict his liability to the Employer to the same extent.

.2 The Contractor must have the Architect's written approval of any restrictions of the NSup's liability prior to entering into a contract with that NSup.

.3 All nominations of Suppliers by the Architect must comply with Clause 36.4.

Remember - the provisions of the Unfair Contract Terms Act, 1977 regarding liability restrictions - exclusion or exemption clauses, particularly related to 'consumer sales'.

Note: Exemption clauses, even if complying with requirements of notice and reasonableness, normally are constructed by the courts strictly against the parties relying on them.

Ailsa Craig Fishing Co. Ltd. v. Malvern Fishing Co. Ltd. & Another (1983), differentiates clauses excluding liabilities from clauses limiting liability - the latter are viewed with less 'hostility' by the courts.

If an NSup wishes to restrict his liabilities or obligations under a proposed supply Contract, the Contractor must obtain the Architect's written consent to the restrictions or else the Contractor himself assumes the liabilities.

Note: Unlike the provisions relating to NS/Cs, the Contractor has *no* right of 'reasonable objection' to an NSup.

Gloucestershire County Council v. Richardson (1968) (House of Lords): The ruling in this case that, due to nomination, the Supplier's exclusion clauses were well known to the Architect (and, by implications of agency, the Employer) who then nominated and thereby gave the Contractor no option but to go to that Supplier, the implied warranty regarding fitness for purpose and latent and patent defects was also excluded from the Contract between the Employer and Contractor in this regard. The exclusion clause was validly passed on.

As indicated above regarding terms of the contract of sale, the Standard Form now stipulates what must be contained in a contract of sale and how liability exclusions are to be dealt with and possibly passed on. The fact of nomination by the Employer's expert agent, the Architect, may be sufficient to overcome the problems of passing on such liability through the operation of the Unfair Contract Terms Act, 1977.

CONDITIONS: PART 3: FLUCTUATIONS

Clause 37: Choice of fluctuations provisions—entry in Appendix

See also Practice Note 17.

37.1 The chosen fluctuations provisions applicable to the Contract must be stated in the Appendix.

The choice is:

Clause 38 - Contribution, Levy and Tax Fluctuations
or
Clause 39 - Labour and Materials Cost and Tax Fluctuations
or
Clause 40 - Use of Price Adjustment Formulae.

37.2 If neither Clause 39 or 40 is identified in the Appendix, Clause 38 applies.

Thus, unless the Appendix provides for the Contract to be executed on the basis of 'full' fluctuations, 'partial' fluctuations provisions are applicable (often termed firm price Contract).

The three fluctuations Clauses are contained in a separate booklet detailing the various versions of each Clause applicable to the Private Edition (separate booklet for L.A. Edition).

No version of Clause 38 exists for the Approximate Quantities Contract and no version of Clause 40 exists for the Without Quantities Contract.

Clause 38: Contribution, levy and tax fluctuations

See also Practice Notes 7 and 17.

Clause 38 is not suitable for use with the Approximate Quantities Edition.

38.1 The Contract Sum is deemed to have been calculated as detailed by this Clause (whether it is or not is a matter for decision by the Contractor at the tendering stage).

Note: at this point it is useful to consider some definitions as contained in Clause 38.6 (also 39.7).

38.6.1 (39.7.1) - 'Date of Tender': the date 10 days before the date fixed for the receipt of Tenders by the Employer.

38.6.2 (39.7.2) - 'Materials' and 'Goods': exclude consumable stores, plant and machinery, but include:

(a) timber used in formwork
(b) electricity
(c) fuels, where specifically so stated in the Contract Bills - see Clause 38.2 (39.3).

38.6.3 (39.7.3) - 'Workpeople': persons whose rates of wages, etc., are governed by the National Joint Council for the Building Industry or some other wage-fixing body for trades associated with the building industry.

38.6.4 (39.7.4) - 'Wage-fixing Body': a body which lays down recognised terms and conditions of workers as under the Employment Protection Act 1975, S11, Paragraph 2(a).

The definitions are applicable to Clause 38 (39) only.

.1 The prices in the Contract Bills are based upon the types ('tender type') and rates ('tender rate') of contribution, levy and tax payable by a person in his capacity as an employer and which are payable by the Contractor at the Date of Tender.

.2 If any tender types or tender rates change, are deleted or new ones are introduced after the Date of Tender, the nett alteration must be paid or allowed by the Contractor.

This applies in respect of:

.1 workpeople on site, and
.2 the Contractor's workpeople off site but working upon or in connection with the Contract (e.g. production of goods - joinery, etc.)

Note: both cases - the operatives are the Contractor's direct employees only.

Note: levies, etc., under the Industrial Training Act, 1964 (C.I.T.B. levy and payment) are expressly

.3 Other direct employees of the Contractor working on or in connection with the Works are, for the purposes of this Clause, treated as craft operatives (as prescribed by Clause 38.1.4).

.4 In such instances, however, the following provisos apply.

(a) Each employee must have worked on or in connection with the Contract for at least two working days during the week against which the claim is applicable. Time aggregation is allowed in respect of whole working days only.

(b) The highest properly fixed craft operative's rate must be used, provided such a craft operative is employed by the Contractor (or Domestic S/C).

(c) The Clause is applicable to those employed by the

Contractor as defined by the Income Tax (Employment) Regulations, 1973, (the P.A.Y.E. Regulations) under S204 of the Income and Corporation Taxes Act 1970.

.5 'Tender type' and 'tender rate' are re-defined for Clause 38.1.6 in respect of refunds to the Contractor in his role as an employer of labour. The definition is extended to include premiums receivable by an employer of labour. Prices in the B.Q. are based on tender types and rates as at the Date of Tender.

.6 The nett alteration to any tender types or rates from those applicable at the Date of Tender must be paid to or allowed by the Contractor.

.7 Premiums are defined as (see Clauses 38.1.5 and 38.1.6):

'...all payments howsoever they are described which are made under or by virtue of an Act of Parliament to a person in his capacity as an employer and which affect the cost to an employer of having persons in his employment.'

.8 Any direct operatives who are 'contracted out' (Social Security Pensions Act, 1975) are deemed not to be so for the purposes of calculating fluctuations in respect of employers' contributions.

Thus, no employee may be contracted out for the calculation of employers' contributions (N.I. contributions).

.9 Contributions, levies and taxes are defined for the purposes of this Clause in a very similar manner to the above definition of premiums - what an employer must pay under Acts of Parliament as an employer of labour.

Note: This will include statutory insurances against personal injury and death.

38.2 The Contract Sum is deemed to have been calculated in the prescribed manner, in respect of materials, goods and fuels, and is subject to adjustments as detailed.

.1 Prices in the Contract Bills are based upon duty and tax types ('tender type') and rates ('tender rate') applicable at the Date of Tender on import, sale, purchase, appropriation, processing or use (except V.A.T. reclaimable as the Contractor's input tax) on

(a) materials
(b) goods
(c) electricity
(d) fuels - if so specifically stated in the B.Q. and specified on a list completed by the Contractor and attached to the B.Q.

This is provided the duty or tax is applicable under an Act of Parliament.

.2 If any tender type or rate is altered from that applicable at the Date of Tender (including electricity and, if applicable, fuel for the temporary site installations) or a deletion or new type or rate occurs, the nett difference between that paid by the Contractor and what he otherwise would have paid must be paid to or allowed by him (except, of course, V.A.T. input tax).

38.3 Domestic Sub-Contractors' items

.1 The Domestic Sub-Contract must incorporate provisions to the same effect as those of Clause 38 of the main Contract together with any appropriate percentage (Clause 38.7).

.2 Any adjustment to the Domestic Sub-Contract sum due to the operation of the Clause 38.3.1 must be passed on to the Employer by the Contractor.

38.4 .1 The Contractor must give written notice to the Architect of the occurrence of any events regarding the provisions of:

.1 Clause 38.1.2 - tender types and rates and workpeople

.2 Clause 38.1.6 - tender types and rates and refunds

.3 Clause 38.2.2 - tender types and rates and materials, etc.

.4 Clause 38.3.2 - Domestic Sub-Contractors' fluctuations.

.2 The written notice is a condition precedent to any payment to the Contractor under this Clause and so must be given within a reasonable time from the occurrence of the event to which it relates.

Thus, a notice is required each time a fluctuation event occurs, although one notice may properly cover several events which occurred at or about the same time.

.3 The Q.S. and Contractor may agree the nett amount of each fluctuation due to each notified event.

.4 Any amount of fluctuations shall form an adjustment to:

.1 the Contract Sum, *and*

.2 any determination payments (Clause 28.2.2.1 and 28.2.2.2). This is subject to the provisions of Clauses 38.4.5, 38.4.6 and 38.4.7.

.5 The Contractor must submit to the Q.S. or Architect any evidence they may reasonably require to calculate the amounts of fluctuations. This must be done as soon as reasonably practical. Where a fluctuation is claimed in respect of employees other than 'workpeople' (not operatives - Clause 38.1.3) the evidence must include a certificate signed by or on behalf of the Contractor each week certifying the validity of the evidence. (This is also applicable to Domestic Sub-Contractors' fluctuations.)

.6 Fluctuations adjustments must never alter the amount of Contractor's profit included in the Contract Sum. Thus fluctuations must be paid nett.

This seems to suggest the sum constituting the Contractor's profit must not be altered by fluctuations adjustments (as the percentage on cost would be).

.7 Fluctuations are not adjustable in respect of payments to the Contractor after the Completion Date, as appropriately amended - the fluctuations are 'frozen' at the Completion Date.

Note: this amends the ruling of Salmon L. J. in Peak Construction (Liverpool) Ltd. v. McKinney Foundations Ltd. (1971).

.8 Clause 38.4.7 applies only if:

.1 Clause 25 applies unamended (Extension of Time), *and*

.2 the Architect has made an award in respect of every written Extension of Time notification (Clause 25).

38.5 Fluctuation provisions are not applicable in respect of:

.1 Dayworks (Clause 13.5.4)

.2 NS/Cs and Nominated Suppliers (but are included in the NSC or Contract of Sale)

.3 Work for which the Contractor's tender under Clause 35.2 has been accepted - Contractor acting as an NS/C also. (Clause 35 tender conditions to apply.)

.4 V.A.T. changes.

38.6 Definitions as detailed above - see under Clause 38.1.

38.7 The percentage stated in the Appendix must be added to fluctuations paid to or allowed by the Contractor under Clauses:

.1 38.1.2

.2 38.1.3

.3 38.1.6

.4 38.2.2

Note: Following J. Murphy Ltd. v. London Borough of Southwark (1981) the term 'workpeople' in the fluctuations clauses does not include labour only sub-contractors.

Clause 39: Labour and materials cost and tax fluctuations

See also Practice Note 17.

As this is the full, traditional fluctuations clause, much of the Clause 38 provisions are reproduced - see Clauses 38.6 and 39.7 - the definitions.

Again, there are several distinct sections.

Clause 39.1 - wages
Clause 39.2 - labour taxes
Clause 39.3 - materials prices (including taxes)
Clause 39.4 - Domestic Sub-Contractors
Clause 39.5 - fluctuations calculations
Clause 39.6 - work not applicable
Clause 39.7 - definitions

39.1 The Contract Sum is deemed to have been calculated as detailed by this Clause and is subject to adjustment as specified.

.1 The prices in the Contract Bills are based upon rates of wages and the other emoluments (payments) and expenses which will be payable by the Contractor. This includes:

(a) Employer's liability insurance (Contractor as an employer of labour), and
(b) Third party insurance, and
(c) Holiday credits.

The payments are in respect of

.1 Workpeople on site, *and*

.2 Contractor's workpeople off site but working on or in connection with the Contract.

The payments must be in accordance with

.3 the rules or decisions of the N.J.C.B.I., or other appropriate wage-fixing body, as applicable to the Works and which have been promulgated at the Date of Tender.

Note: promulgated - published as coming into force or having authority. Thus any wage changes which have been agreed and published as at the Date of Tender, even if not operative until some time after that date, are deemed to have been taken into account in the calculation of the Contract Sum and are, therefore, not subject to the fluctuations provisions.

.4 any incentive scheme/productivity agreement as advised and recognised by N.J.C.B.I. (Working Rule Agreement 1.16, or its successor, including the general principles provisions of the Working Rule Agreement) or other appropriate body.

.5 the terms of the Building and Civil Engineering Annual and Public Holiday Agreements (of N.J.C.B.I. or other appropriate body) as applicable to the Works and which have been promulgated at the Date of Tender.

The prices in the B.Q. are deemed to be based also upon the rates or amounts of any contribution, levy or tax payable by the Contractor as an employer in respect of (or calculated by reference to):

(a) rates of wages, and
(b) other emoluments, and
(c) expenses, including holiday credits as specified in this Clause.

Note: By the provisions of Clause 39.2.2, the contributions and levies of the C.I.T.B. etc. are expressly excluded from the fluctuations provisions.

.2 If any rules, decisions or agreements promulgated after the Date of Tender alter the rates of wages, other emoluments and expenses, the nett amount of the change must be paid to or allowed by the Contractor.

Such adjustment is inclusive of any consequential change in the cost of

(a) Employer's liability insurance, and
(b) third party insurance, and
(c) any contribution, levy or tax payable by a person as an employer of labour.

Thus, if wage rates increase after the Date of Tender due to a promulgation after that date, both that increase and any consequential increase in, say, employer's liability insurance are recoverable by the Contractor. (Employer's liability insurance is often based upon the wages bill of a firm.)

However, if the Employer's liability insurance premium is increased due to (a) a wage increase and (b) inflation generally, then only (a) is recoverable under this Clause.

To be recoverable, the consequential increase must be directly and solely due to one or more of the specified causes (increase in rates of wages on the Contract), not just more expensive insurance.

Despite the House of Lords ruling under the 1963 Form in William Sindall Ltd. v. N. W. Thames Regional Health Authority (1977), it would now appear from the wording of this Clause 39 that increases in the cost of productivity bonuses directly resulting from increases in standard rates of wages (established by N.J.C.B.I. or other appropriate wage-fixing body) are recoverable. The applicable proviso is that the bonus scheme is:

(a) covered by the Working Rule Agreement, as required by Clause 39.1.1.4, *and*
(b) recognised by the unions.

It is suggested that a casual, unrecognised, site agreement would be outside the scope of this Clause and so the Sindall ruling would apply:

(a) the consequential bonus increase is not recoverable,

but

(b) if a productivity or other bonus scheme is being operated 'casually', although consequential increases in these payments are not recoverable, increases in guaranteed minimum bonus are recoverable - they represent guaranteed minimum payments.

.3 Contractor's employees, outside the definition of workpeople, engaged on or in connection with the Works, are subject to fluctuations recovery as if they were craft operatives.

.4 (This is as Clause 38.1.4 but is used here to cover wage rate increases.)

Note: The wording of the Clause implies that production bonus, other than guaranteed minimum, payments should be ignored when calculating fluctuations in respect of these employees.

.5 Fares, etc.

The prices contained in the B.Q. are based upon:

(a) transport charges incurred by the Contractor (as set out in the basic transport charges list attached by the Contractor to the B.Q.) in respect of workpeople - Clauses 39.1.1.1 and 39.1.1.2, *or*

(b) reimbursement of fares to workpeople (Clauses 39.1.1.1 and 39.1.1.2) under the rules of N.J.C.B.I. (or other appropriate body) promulgated at the Date of Tender.

.6 The nett amount of any changes in cost of fares, etc., to the Contractor, as defined above, must be paid to or allowed by him on the basis of:

.1 changes in transport charges set out in the basic list of transport charges occurring after the Date of Tender

.2 actual reimbursement taking effect or promulgation of changes (as appropriate) after the Date of Tender.

39.2 Largely a reproduction of parts of Clause 38.

.1 Clause 38.1.1

.2 Effectively Clause 38.1.2

.3 Clause 38.1.3 incorporating Clauses 38.1.4 and 39.1.4

.4 Clause 38.1.5

.5 Clause 38.1.6

.6 Clause 38.1.7

.7 Clause 38.1.8 except where contracted out employee pension schemes are under the rules of N.J.C.B.I., or other appropriate body, such that contributions are covered by the provisions of Clause 39.1 and are thereby changes in contributions and recoverable.

.8 Clause 38.1.9.

39.3 Materials - The Contract Sum is deemed to have been calculated as specified and is subject to adjustment accordingly.

.1 The prices contained in the B.Q. are based upon the market prices of materials, goods, electricity and, where specifically so stated in the B.Q., fuel - list required (as Clause 38.2.1). The market prices must have been current at the Date of Tender and are termed 'basic prices'. The prices set out by the Contractor in the list (basic price list) and attached to the B.Q. are deemed to be the basic prices of the items therein specified.

Note: Electricity and any fuels must be consumed on site for the execution of the Works, including temporary site installations.

.2 Any changes in the market prices of items specified on the basic list (including electricity and fuels for the Works and temporary site installations) occurring after the Date of Tender are the subject of fluctuations adjustments of the nett changes.

.3 Market prices include any tax or duty (except V.A.T. input tax of the Contractor) on the import, purchase, sale, appropriation, processing or use of goods, materials, electricity or fuels as specified, provided such tax or duty is governed by an Act of Parliament.

"Market Price" - Lord Kinnear in Charrington & Co. Ltd. v. Wooder (1914):

"I am unable to accept the contention that the term 'market price' has a fixed and definite meaning which must attach to it invariably, in whatever contract it may occur, irrespectively of the context or the surrounding circumstances. The argument was rested chiefly on the force which, it is said, must be given to the work 'market'. In a different connection this may be a technical term, but the covenant in question is not used in any technical sense, and in ordinary language it is a common word, of the most general import. It may mean a place set apart for trading, it may mean simply purchase and sale, there are innumerable markets each with its own customs and conditions. Words of this kind must vary in their signification with the particular objects to which the language is directed; and it follows that a contract about a market price cannot be correctly interpreted or applied without reference to the facts to which the contract relates."

Thus, market price must be viewed in context. If there is only one supplier, that supplier's price is the market price.

If there are many suppliers the market price will have a range, the Contractor's usual supplier's price indicating the market price for that Contractor.

If an item becomes restricted in the suppliers from whom it is obtainable, the Contractor must obtain that item from one of these suppliers, possibly at an elevated price but that is the market price of the item to the Contractor at that time.

Usually Contractors obtain commodities from the suppliers whose quotations were used in the preparation of the basic price list.

39.4 Work let to Domestic Sub-Contractors - corresponding provisions to those of Clause 38.3 but of increased scope to cover wage fluctuations.

39.5 As Clause 38.4 but the notice to be in respect of

.1.1 Clause 39.1.2 - wages

.2 Clause 39.1.6 - transport and fares

.3 Clause 39.2.2 - tender types and tender rates - alterations

.4 Clause 39.2.5 - tender types and tender rates - alterations/deletions/additions

.5 Clause 39.3.2 - market prices

.6 Clause 39.4.2 - Domestic Sub-Contractors

.2 Clause 38.4.2

.3 Clause 38.4.3

.4 Clause 38.4.4

.5 Clause 38.4.5

.6 Clause 38.4.6

.7 Clause 38.4.7

.8 Clause 38.4.8.

39.6 Work not subject to fluctuations as Clause 38.5.

39.7 Definitions - as Clause 38.6.

39.8 .1 Percentage addition to be added, as stated in the Appendix, to fluctuations under:

.1 Clause 39.1.2

.2 Clause 39.1.3

.3 Clause 39.1.6

.4 Clause 39.2.2

.5 Clause 39.2.5

.6 Clause 39.3.2.

In respect of the provisions of Clause 39.1.5, transport charges, and 39.3.1, materials including electricity and fuels, it would be of value if Q.S.s were to provide suitably headed sheets on which the Contractor should enter his items to comprise the basic lists for fluctuations recovery and to submit these with his tender or Bills, as required.

Clause 40: Use of price adjustment formulae

See also Practice Note 17.

The formula method of fluctuations adjustments is rapidly becoming more widely accepted and used. It provides for full fluctuations recovery and is often termed the N.E.D.O. Formula, after the indices upon which it is based. There is no provision for the use of formulae fluctuations in the Without Quantities Editions.

The adjustment applies to all payments *except* releases of Retention as these will have already been the subject of formula adjustments.

Note: Clause 30.2.1.1 provides for Retention to be deducted from formula fluctuations adjustments.

40.1 .1.1 Prescribes for the Contract Sum to be adjusted in accordance with Clause 40 and the J.C.T.'s Formula Rules current at the Date of Tender.

.2 Any formula adjustment is exclusive of V.A.T. and must not affect the operation of Clause 15 - V.A.T. Supplemental Provisions and the V.A.T. Agreement.

.2 The Definitions in Rule 3 of the Formula Rules are to apply to Clause 40.

.3 The adjustments are to be effected in all certificates

for payment. (L.A. Forms require any non-adjustable element to be deducted prior to the adjustment being made.)

.4 If any adjustment correction is required (Rule 5), it must be effected in the next certificate for payment.

40.2 Interim Valutions *must* be made before the issue of each Interim Certificate. Clause 30.1.2 is deemed amended accordingly.

40.3 Articles manufactured outside the U.K. (Rule 4(ii)) - Contractor to append a list to the B.Q. giving the sterling market price of each item (delivered to site) as at the Date of Tender.

Any variation in the market price is subject to adjustment under similar provisions to those of Clause 39, i.e. exclusive of Contractor's input V.A.T.

40.4 Nominated Sub-Contractors

.1 Method of adjustment of the Nominated Sub Contract Sums to be agreed, tendered upon and approved in writing by the Architect prior to the issue of the nominating A.I.s.

The available methods are:

.1 Electrical installations, Heating and Ventilating and Air Conditioning installations, Lift installations, Structural Steelwork installations and Catering Equipment installations. The relevant specialist formula must be used - Rules 50, 54, 58, 63, 69, *or*

.2 where none of the specialist rules applies:

(a) Formula as Part 1 of Section 2 of the Rules and

(b) One or more of the Work Categories as set out in Appendix A to the Rules, *or*

.3 some other method if the other possibilities are inapplicable.

.2 Domestic Sub-Contracts
The Sub-Contracts must provide for the Sub-Contract Sums to be adjusted as follows, subject to agreement

between the Contractor and Domestic Sub-Contractor of an alternative method

.1 as Clause 40.4.1.1

.2 as Clause 40.4.1.2

40.5 The Q.S. and Contractor may agree any alteration to the methods and procedures for the formula fluctuations recovery. Any amounts of adjustments so determined are deemed to be the formula adjustment amounts.

This is subject to:

.1 the amounts of adjustment calculated after amendment must approximate to those obtained by operation of the Rules as laid down, *and*

.2 any such amendments must not affect the provisions relating to Sub-Contractors (as Clause 40.4).

40.6 .1 Prior to the issue of the Final Certificate, if the publication of the Monthly Bulletins is delayed or ceases thereby preventing the formula adjustment being properly effected, adjustment is to be made on a fair and reasonable basis to determine the amount due in each Interim Certificate so affected.

.2 Prior to the issue of the Final Certificate, if the publication of the Monthly Bulletins is recommenced, Clause 40 and the Formula Rules are to apply retrospectively. The index adjustment from the Bulletins must be carried out and will prevail over the fair adjustments made.

.3 During any cessation of publication of the Monthly Bulletins, the Contractor and Employer must operate the relevant provisions of Clause 40 and the Formula Rules so that on publication being recommenced the appropriate formula adjustments may be made retrospectively.

40.7 .1.1 If the Contractor does not complete the Works by the Completion Date, as applicable:
the value of work completed after that date is subject to formula adjustment on the basis of the indices applicable at the relevant Completion Date.

Thus items completed in a time over-run are adjusted under the formula method as if they were completed in the month prior to the applicable Completion Date.

.2 If items completed in an over-run period are adjusted in any other way, such adjustment must be corrected to accord with that specified by Clause 40.7.1.1, as above.

.2 Clause 40.7.1 does *not* apply *unless*:

.1 Clause 25 (Extension of Time) applies, unamended, to the Contract, *and*

.2 the Architect has fixed the Completion Date as appropriate, in writing, in respect of every written notification by the Contractor under Clause 25.

Appendix

The provisions of the Appendix are quite self-explanatory and have been referred to throughout the consideration of the Conditions.

Mostly the Appendix provides for insertions by the Employer (and/or consultants on his behalf) where the Contract makes this appropriate e.g. liquidated and ascertained damages - Clause 24.2 - at the rate of £_________ per ________.

The Appendix therefore provides a considerable amount of information about the particular Contract in a very concise way. Much of the information is non-standard. Thus, the Appendix is of great importance to the parties (and consultants) and must always be examined and analysed with great care to determine the full implications (this is notably vital at tender stage).

Supplemental Provisions (the VAT Agreement)

See also Practice Note 6.

The Supplemental Provisions are incorporated into the Contract by Clause 15.1.

1 The Employer must pay to the Contractor, in the manner specified, any V.A.T. chargeable by the Commissioners (of Customs and Excise) on the Contractor in respect of the supply of goods and services to the Employer under the Contract.
(Under any current amendments of Regulation 21(.1) of the Value Added Tax (General) Regulations, 1972).

Thus, the Employer must pay the Contractor the appropriate amount in respect of any positively rated items in the Contract, at the stipulated rate(s).

.1 The Contractor must give the Employer a written provisional assessment of the values (less any applicable Retention) of supplies of goods and services included in a payment Certificate which are subject to V.A.T. (i.e. all *except exempt* items).

The timing of such assessment must be:

(a) not later than the date for the issue of each Interim Certificate

(b) not later than the date for the issue of the Final Certificate, unless the procedure detailed in Clause

1.3 has been completed.

The assessment must show the supplies:

.1 subject to zero rate of tax (Category (i)), *and*

.2 subject to any rate(s) of tax other than zero (Category (ii)), i.e. positively rated.

In respect of any Category (ii) items, the Contractor must state:

(a) the rate(s) applicable, *and*
(b) the grounds upon which he considers the supplies so chargeable.

.2.1 On receipt of a written provisional assessment, the Employer must (unless he (may reasonably) objects to the assessment):

(a) calculate the amount of tax due on the basis of the details of the Category (ii) items,
(b) pay that amount to the Contractor within the period for honouring (together with the amount due ordinarily under the Interim Certificate).

.2 If the Employer has reasonable grounds for objection to the provisional assessment, he must so notify the Contractor in writing, specifying the grounds within 3 working days of receipt of the assessment. Within 3 working days of receipt by the Contractor of the Employer's notice, he must reply in writing either:

(a) withdrawing his assessment and thereby releasing the Employer from the obligations to calculate and pay tax in respect of the assessment - Clause 1.2.1, *or*
(b) confirm the assessment. The Contractor then regards any amount received from the Employer in connection with the assessment and associated Certificate as being inclusive of the appropriate V.A.T. (i.e. a gross payment) and issues an authenticated receipt accordingly - Clause 1.4.

.3.1 As soon as possible after the issue of the Certificate of Completion of Making Good Defects, the Contractor must prepare a final written statement of the respective values of all supplies of goods and services for which certificates have been or will be issued and which are chargeable on the Contractor at

.1 zero rate - Category (i), and

.2 any rate(s) of tax other than zero - Category (ii)

The final statement must be issued to the Employer. As with the interim assessments, in respect of Category (ii) items the Contractor must specify:

(a) the rate(s) applicable, *and*
(b) the grounds upon which he considers the supplies so chargeable, *and*, for the final statement only,
(c) the total amount of tax already received by the Contractor for which receipt(s) have been issued - as Clause 1.4 of these provisions.

.2 The final statement may be issued either before or after the issue of the Final Certificate. Practically, it is probably advantageous to issue the statement after the issue of the Final Certificate.

.3 The Employer must, on receipt of the final statement, calculate the final amount of tax due upon the details of the statement. He must pay any balance of tax (total per statement calculations less tax already paid) to the Contractor within 28 days from receipt of the statement.

.4 If the Employer discovers the amount of tax in accordance with the final statement is less than the amount he has already paid, he must so notify the Contractor. The Contractor then must:

(a) refund the excess to the Employer within 28 days of receipt of the notification, *and*
(b) accompany the refund by a receipt (Clause 1.4 of the provisions) showing the correction of previous receipt(s).

.4 The Contractor must issue receipts upon receiving monies under certificates and the appropriate amount of tax (Clause 1).

The receipts must comply with Regulation 21(2) of the V.A.T. (General) Regulations, 1972, including the particulars as required by Regulation 9(1), taking account of any amendments and/or re-enactments.

2 .1 The Employer must disregard any set-off in respect of liquidated damages when calculating and paying amounts of V.A.T. due to the Contractor (Clauses 1.2 and 1.3 of these provisions).

.2 The Contractor must likewise ignore contra-charges of liquidated damages by the Employer in the preparation of the final statement.

3 .1 If the Employer disagrees with the Contractor's final statement he may ask the Contractor to obtain the decision of the Commissioners on the tax properly chargeable. This request must be made before the tax payment (or refund) becomes due.

If the Employer then disagrees with the Commissioners' decision he must request the Contractor to make such appeals to the Commissioners as he instructs. In such instances the Employer must indemnify the Contractor against all costs and other expenses. (The Contractor also has an option to be secured by the Employer against such costs and expenses.)

The Contractor must account to the Employer for any costs awarded in his favour in any appeals under Clause 3 of the provisions.

.2 If required to do so, before an appeal may proceed, the Employer must pay to the Contractor the full amount of tax alleged to be chargeable.

.3 The balance of tax must be paid by or refunded to the Employer within 28 days of the final adjudication of an appeal, or from the date of the Commissioners' decision if no appeal is to be made.

Authenticated receipts must also be issued accordingly under Clause 1.3.4 of these provisions.

4 The Employer is discharged from further liability to pay tax to the Contractor upon settlement, as prescribed, of the amount of the final statement, Commissioners' decision or appeal decision.

The exception is if the Commissioners introduce a correction to the tax charged subsequently. (This is also subject to the prescribed appeal provisions of Clause 3 of these provisions.) Thereupon the Employer must pay the additional amount to the Contractor.

5 Awards of an Arbitrator or Court which vary payments between the parties. Such alterations in payments must also be subject to V.A.T. as applicable.

6 Arbitration is not applicable to V.A.T. assessments by the Commissioners (Clause 3 of these provisions).

7 If the Contractor does not provide a receipt, the Employer is not obliged to make any further payments to the Contractor. This applies only if

.1 The Employer shows that he requires the receipt to validate a claim for credit for tax paid or payable under the Agreement which he is entitled to make to the Commissioners, *and*

.2 The Employer has paid tax in accordance with the provisional assessments unless he has sustained a reasonable objection thereto.

8 If the Employer determines the employment of the Contractor (Clause 27.4 of the Conditions), any additional tax which the Employer has had to pay due to the determination may also be set-off by him against any payments to be made to the Contractor (or may be recovered by the Contractor).

Sectional Completion Supplement

The supplement applies to both the Private and Local Authorities Editions, with Quantities.

The Contract Standard Forms do permit partial possession by the Employer - Clause 18. However, that Clause is not intended to be used in such a way that the Contract is, in fact, a phased Contract. In such a case the Sectional Completion Supplement must be used.

The decision is a reflection of the Employer's initial intention - if a phased project, use the Sectional Completion Supplement; if a non-phased project but subsequently the Contractor completes in a 'sectional manner' and the Employer wishes to take possession of the completed section(s), Clause 18 is applicable.

Practice Note 1 is a useful guide to practice and provides a set of amendments to the Contract to facilitate sectional completion.

Contractors must be informed at Tender stage:

(a) The Works are to be carried out in phased Sections.

(b) The Employer will take possession of each Section on Partial Completion of the Section.

(c) The identity of the Sections - by drawings or B.Q. description.

(d) The order and phasing of completion.

(e) Any work common to more than one Section must be a separate Section.

(f) In respect of each Section: Value, Date for Possession, Date for Completion, Rate of Liquidated Damages for delay, Defects Liability Period.

The major adaptations of the Contract involve the division of the Works into distinct Sections with the values ascribed thereto summing to the Contract Sum. Separate completion provisions apply for each Section but only one Final Certificate is issued at the conclusion of the project.

Clause	Adapatation
1.3	defines Section
2.1	Contractor is obliged "to carry out and complete the Works by Sections"
18.1.6	Section value is "the value ascribed to the relevant Section in the Appendix"
	The Section value must be the total value of that Section obtained from the Contract Bills. The Section values must add up to the Contract Sum and must take into account the apportionment of Preliminaries and similar items priced in the Contract Bills.

Procedure for Sectional Completion:

Contractor to be given possession of each Section.

Contractor to execute the Sections successively or concurrently in accordance with the Contract.

Liquidated Damages are calculated and paid separately.

A Practical Completion Certificate must be issued for each Section as Practical Completion is achieved.

Practical Completion of a Section:

(a) relieves the Contractor of any duty to insure that Section under Clause 22A
(b) Clause 30.4 operates to release *approximately* one half of the Retention held in respect of that Section
(c) Clause 17.2 - commences a separate D.L.P. for that Section.

When all Sections are complete, the Architect must so certify — the period for the Final Account and Final Certificate for the whole Works commences at the date of that Certificate (Clause 30.6).

Insurance arrangements under Clause 21.2 - amount of insurance to be maintained in respect of persons and property - must be clarified to determine whether or not any Sections of the Works which have obtained a Certificate of Practical Complètion are then covered by such insurance as "property other than the Works".

The actual changes in the Contract to effect Sectional Completion involve quite minor changes of clause wordings. The changes are, in reality, quite straight-forward and 'common sense' in nature. The amendments to the Appendix provide the greatest amount of information about the requisite changes and in a concise form.

Contractor's Designed Portion Supplemen

See also Practice Note CD/2.

This supplement may be used with both Private and Local Authorities Editions with Quantities.

The supplement provides express terms to incorporate design by the Contractor into the contract. Otherwise the Contractor has no express liability for design but is subject to design liability only to inform the Architect of suspected design defects (see Equitable Debenture Asset Corpn. Ltd. v. William Moss & Others (1984)).

Amendments to be made to the Standard Form are:

(a) replacement of the Recitals and Article 1 by amended versions
(b) modifications to the Conditions
(c) addition of a Supplementary Appendix
(d) addition of the words 'With Contractor's Designed Portion Supplement' at the top of the endorsement on the outside back cover of the Contract Form.

There are also three additional Contract Documents which must be signed by or on behalf of the parties, they are:

(a) the Employer's Requirements for the Contractor's Designed Portion (CDP)
(b) the Contractor's Proposals for the CDP, and
(c) the Analysis of the portion of the Contract Sum to which the CDP relates.

The three additional Contract Documents must be produced as part of the tendering processes - the Employer's Requirements, by or on behalf of the Employer; the Contractor's Proposals and the Analysis, by the Contractor.

The Architect retains overall responsibility for design and is responsible for the integration of the CDP into the Works and to issue directions regarding such integration; the Architect's powers in this regard are contained in Article 1 and in Clause 2.1.3.

The Second Recital identifies the CDP as part of the Works (the latter being denoted in the First Recital).

The Third Recital identifies who has prepared the drawings etc., including the Employer's Requirements.

The Fourth Recital requires that the Contractor has submitted 'the Contractor's Proposals' for the design and construction of the CDP and 'the Analysis' of the part of the Contract Sum relating to the CDP.

Note: Contents and format of the Analysis are at the discretion of the Contractor unless requirements therefore, such as those outlined in Practice Note CD/2, are provided in the tender documents - particularly the Bills of Quantities.

The Seventh Recital stipulates that the Employer has examined the Contractor's Proposals in the context of the Conditions and is satisfied that the proposals appear to meet the Employer's Requirements.

Note: The Recital indicates acceptance in principle of the proposals but seeks to ensure that the Contractor retains responsibility for their actually meeting the Employer's Requirements.

Other Recitals are almost identical in content to those of JCT '80 but incorporate appropriate amendments to allow for the CDP. There are eight Recitals in the CDP Supplement.

Article 1, Contractor's Obligations incorporate provision for the Contractor to complete the design of the CDP and to do so as the Architect may direct in order that the design of the CDP may be integrated with the design for the Works. The design obligation is in addition to the obligation for the Contractor to carry out and to complete the Works as per the Contract Documents, including the modifications thereto specified within the CDP Supplement.

The remainder of the CDP Supplement details modifications to be made to the Conditions. Many of the modifications are straightforward and are necessary to incorporate the additional documents, definitions etc.

Some of the more important amendments are noted below:

Clause	Adaptation
2.1.2	Extends the scope of 'Contract Documents'.
2.1.3	Contractor to design the CDP and to incorporate all specifications etc. not noted in the Employer's Requirements but which are necessary to achieve completion of the Works. Architect may give directions re. design of CDP for integration with the total design for the Works.
2.1.4	Where the Architect is required to approve standards of materials or work, the standards are to be to the Architect's reasonable satisfaction.
2.2.2.3	Errors of description, quantity or omission from the Contractor's Proposals and the Analysis must be corrected with no change to the Contract Sum; this includes errors etc. which necessitate an A.I. constituting a Variation for their correction.
2.4.1	Discrepancies, divergencies between Contractor's Proposals, Analysis and other design documents prepared by the Contractor - Contractor to notify the Architect of the discrepancy/divergence and, as soon as possible, to provide a statement of proposed amendments to overcome the problem. Once the Architect has received the statement, the obligation for the Architect to issue instructions under Clause 2.3 comes into effect.
2.4.2	A.I.s given following Clause 2.4.1 shall not occasion any addition to the Contract Sum.
2.5	Provision of supplementary documents, levels etc. to amplify the Contractor's Proposals by the Contractor to the Architect without change.
2.6.1	Contractor assumes design liability for the CDP to the same level as an Architect or appropriate professional designer holding themselves out to be competent to execute the design. Note: The standard of care which the Contractor must assume is that of an expert.

See Bolam v. Friern Barnet Hospital (1957) and Greaves v. Baynham Meikle (1975).

2.6.2 Passes on to the Contractor responsibility under Clause 2.6.1 for compliance with Section 2(i) of the Defective Premises Act, 1972 where the CDP concerns work in connection with a dwelling or dwellings and the Employer's Requirements refer to such liability.

2.6.3 Where the Contract does not involve the Contractor working in dwellings under the Defective Premises Act, 1972, the Contractor's liability for other parties' losses arising out of the Contractor's failure under Clause 2.6.1 is limited to the amount if any as stated in the Supplementary Appendix. This Clause has no bearing upon the damages provisions under Clause 24.1.

2.6.4 Contractor's design includes designs prepared by others on behalf of the Contractor.

2.7 If the Contractor believes that compliance with an A.I. or any direction given under Clause 2.1.3 injuriously affects the efficacy of the design of the CDP, the Contractor has 7 days from receipt of the A.I. or direction to so notify the Architect in writing. The notice must specify the injurious effect(s). The A.I. or direction will not take effect unless confirmed by the Architect.

2.8 The Architect must notify to the Contractor any items which appear to the Architect to be defects in design under Clause 2.6.1 (i.e. design defects in the Contractor's Design). Despite any such notices, the Contractor retains full responsibility for the Contractor's Design.

2.9 No extension of time may be awarded under Clause 25.3 nor any loss and/or expenses paid under Clause 26.1 and 28.1.3 in respect of:

.1 errors, divergencies etc. in Contractor's Proposals or supplementary documents provided under Clause 2.5

.2 late provision by the Contractor of drawings etc. concerning the CDP as per Clause 2.5.2

.3 Non receipt in due time by the Architect of drawingsetc. concerning the CDP from the Contractor for which the Architect, specifically applied in writing on a date which having regard to the Completion Date was neither unreasonably distant from, nor unreasonably close to, the date on which it is necessary for him to receive the same.

5.9 Prior to the commencement of the D.L.P., the Contractor shall provide free to the Employer drawings etc., concerning the CDP re. maintenance and operation of that Portion as specified in the Contract Documents or as the Employer reasonably may require.

Note: It would be advisable for the Contractor to provide for giving the Employer a full set of as-built drawings etc. and any available maintenance documents (e.g. maintenance manuals of plant).

6.1.6 If either the Contractor or the Architect finds any divergence between the Contractor's Proposals etc. (Clauses 2.3.5, to 2.3.8) and the Statutory Requirements, he must give the other written notice specifying the divergence. The Contractor must give the Architect written notice of his proposed amendment(s) to remove the divergence; the Architect shall issue instructions to remove the divergence and the Contractor must comply with such A.I.s free of charge. However, if the cause of the divergence is changes in the Statutory Requirements which occurred after the Date of Tender (Clause 6.1.8) and necessitate a change to the C D P , such change shall be treated as a Variation under Clause 13.2.

6.1.7 Emergency work to comply with Statutory Requirements - Contractor to execute only what is reasonably necessary to secure compliance and to inform the Architect of what is being done.

8.1 As far as possible materials goods and work in the C D P to comply with Employer's Requirements. Substitutions require the Architect's written consent. This is in addition to the materials etc. so far as procurable complying with the Contract Bills where not covered by the C D P.

13.2 Architect may issue instruction requiring

a Variation. No Variation....shall vitiate this Contract (see Clause 13 of main text). Architect may sanction in writing any Variation made by the Contractor which is not covered by an A.I. Any A.I. regarding the content of the C D P acts as a statement of amendment to the Employer's Requirements.

13.4.1 Valuation of work done under A.I.s re. expenditure of provisional sums:

by Q.S. as Clause 13.5, but
by Q.S. as Clause 13.8 in respect of C D P Works.

13.8 Valuation rules for Variations to C D P works. Similar rules to valuation of Variations to the main works but:

(a) based on the Analysis instead of the Contract Bills, and
(b) to incorporate adjustment for the addition/ omission of design work.

19.2.2 Written consent of the Employer must be obtained for the Contractor to sub-let any design for the C D P. Consent may not be unreasonably withheld. Contractor retains full responsibility towards the Employer under Clause 2.6.

30.10 Usually no certificate of the Architect will be conclusive evidence that:

.1 any works, materials or goods to which it relates; or
.2 any design to be prepared and completed by the Contractor for the Contractor's Designed Portion,

are in accordance with this Contract.

The Supplementary Appendix identifies:

The Employer's Requirements
The Contractor's Proposals
The Analysis

Clause 2.6.2: whether the scheme is approved under S.2(1) of the Defective Premises Act, 1972.

Clause 2.6: Limit of Contractor's Liability for loss of use etc.:-

Clause 2.6.3 does not apply, or
Clause 2.6.3 applies, with a limit of £_______ (to be inserted, if applicable).

The bulk of the remaining amendements to incorporate the Contractor's Designed Portion Supplement concern modifications to Clause 40 to facilitate operation of the N.E.D.O. price adjustment formulae.

Approximate Quantities Form

See also Practice Note 7.

This Edition is based upon and closely follows the provisions of the With Quantities Form. The main differences are noted below. As the Form uses Approximate Quantities, complete re-measurement of the Works in their 'as executed' form is required.

'Bills of Approximate Quantities' are used throughout, where appropriate, in place of 'Bills of Quantities'.

Recitals: The Bills of Approximate Quantities provide a reasonably accurate forecast of the quantities of work to be executed, the total of the prices in which form the Tender Price.

Article 2:	The total sum to be paid to the Contractor is the Ascertained Final Sum.
Clause 2.3:	Quantity divergencies etc. are dealt with by remeasurement under Clause 14.
Clause 13:	Corresponds to Clause 14 of the With Quantities Form.
Clause 14:	Corresponds to Clause 13 of the With Quantities Form. Variations of quantity of work do not apply except where the quantity was not accurately forecast in the Bills - it is submitted that the concept of reasonableness will be applicable to determine what constitutes an accurate forecast.

Clause 25.4.13: An extra Relevant Event, where the quantity of work was not forecast accurately.

Clause 30.1.2: Valuations are required for each Interim Payment.

Clause 30.6: Notably different from the With Quantities Form. The approach is to build up the final account from 'nothing' by measurement and valuation of the work done and claims applicable under the terms of the Contract.

Clause 38: Not available for use under the Approximate Quantities Form.

Clause 40: Use of Work Groups or a Single Index is inappropriate.

Without Quantities Forms

The Without Quantities Forms are extremely similar to their With Quantities counterparts.

Most of the changes are obviously brought about by the absence of a Bill of Quantities and the presence, in its place, of a Specification and a Schedule of Rates.

Several Clauses are amended solely by the substitution of reference to the "Specification and/or Schedule of Rates" for those to the "Contract Bills".

The more significant alterations are considered below.

Article 4: This is now split into two alternatives:

Article 4A - a reproduction of Article 4, to be used where a Quantity Surveyor will be employed.

Article 4B - a slight change of working from Article 4, to be used where the functions of the Quantity Surveyor are to be carried out by someone who is not a Q.S.

Clause 1.3: The definition of Contract Bills is omitted.

Definitions are included for

(a) Schedule of Rates - Clause 5.3.1.3
(b) Specification - First Recital

Clause 2.2.2: This Clause replaces that in the With Quantities Form which is sub-divided into two sections.

The Clause states that if any errors or omissions exist in the Contract Drawings and/or Specification, they will not vitiate the Contract (but see qualifying notes under the With Quantities form) but must be corrected under the rules of Clause 13.2, as if they were Variations due to Architect's Instructions.

Clause 13.5: This Clause is largely similar to its With Quantities counterpart, except for the following:

Clause 13.5.3 is omitted and the substitution is made:

Clause 13.5.3 now requires that in the valuation of any Variation (Clause 13.5.1 and 13.5.2) appropriate allowances must be included for any consequent changes in Preliminaries (as defined by S.M.M.6).

Appendix – Brief Revision Guide

This section provides a list of the contents of the Contract. The information is tabulated in four columns;

Column 1 - Article or Clause number
Column 2 - Article or Clause title - indicating the provisions
Column 3 - Note of the persons primarily concerned with the operation of the provisions
Column 4 - Brief summary of the main provisions.

It should be noted that everyone connected with the operation of the Contract should be aware of all the provisions. Particularly the Employer and Contractor will be affected by every provision. The purpose of column 3, therefore, is to denote those persons most usually and directly affected by the operation of that part of the Contract.

The following abbreviations have been used:

Arch.	Architect
B.I.	Building Inspector
Contr.	Contractor
C.o.W.	Clerk of Works
DS/C	Domestic Sub-Contractor
Empl.	Employer
Govt.	Government
L.A.	Local Authority
NS/C	Nominated Sub-Contractor
NSup	Nominated Supplier
Q.S.	Quantity Surveyor
R.I.B.A.	Royal Institute of British Architects
S.U.	Statutory Undertaker.

REF. NO.	SUBJECT	PERSONS PRIMARILY CONCERNED	MAJOR CONTENTS/ COMMENTS
ARTICLES OF AGREEMENT			
1	Contractor's obligations	Contractor	'Blanket' provision - execute prescribed work
2	Contract Sum		Subject only to express modifications - fluctuations
3	Architect	Architect	Employer's agent for express purposes - condition
4	Quantity Surveyor	Q.S.	Must value variations. Must execute interim valuations if Clause 40 applies.
5	Settlement of Disputes	Contr. Empl. Arch. Q.S. CoW. RIBA?	Alternative to action at Law. Usually after Practical Completion
CONDITIONS: PART 1 - GENERAL			
1	Interpretation, definitions etc.	All	
1.1	Method of reference to clauses	All	As with any legal document
1.2	Articles etc. to be read as a whole	All	
1.3	Definitions		
2	Contractor's obligations		
2.1	Contract documents	All	Drawings, Bills, Articles, Conditions, Appendix
2.2.1	Contract Bills - relation to Articles, Conditions and Appendix	Q.S.	Bills do not over-ride
2.2.2	Preparation of Contract Bills - errors in preparation etc.	Q.S.	Corrected as a Variation
2.3	Discrepancies in or divergences between documents	Arch. Q.S. Contr.	A.I. required - Contractor's request
3	Contract Sum - additions or deductions - adjustment - Interim Certificates	Contr. Arch. Q.S.	Adjusted a.s.a.p.
4	Architect's instructions	Contr. Arch	
4.1	Compliance with Architect's inst's		Immediate unless objection; express authority for A.I.
4.2	Provisions empowering instructions	Arch.	Architect may specify provision allowing A.I.
4.3.1	Instructions to be in writing	Arch.	
4.3.2	Procedure if instructions given otherwise than in writing	Arch. Contr.	Confirmation by Contractor or Architect
5	Contract documents - other documents - issue of certificates		
5.1	Custody of Contract Bills and Contract Drawings	Arch. Q.S. Contr.	Architect or Q.S., Contractor may inspect

REF. NO.	SUBJECT.	PERSONS PRIMARILY CONCERNED	MAJOR CONTENTS/ COMMENTS
5.2	Copies of documents	Contr. Arch.	Contractor's entitlement
5.3	Descriptive schedules etc. - master programme of Contractor	Contr. Arch.	Bills should specify any required master programme
5.4	Drawings or details	Arch. Contr.	2 free copies for Contractor
5.5	Availability of certain documents	Contr.	
5.6	Return of drawings etc.	Arch. Contr.	Property of Architect
5.7	Limits to use of documents	All	For the project only
5.8	Issue of Architect's certificates	Contr. (Q.S.)	
6	Statutory obligations, notices, fees and charges	B.I. etc.	
6.1	Statutory Requirements	Contr. Arch.	Contractor must comply; discrepancy A.I? Emergencies
6.2	Fees or charges	Contr. Q.S. Empl.	Usually Contractor's responsibility
6.3	Exclusion of provisions on Domestic and Nominated Sub-Contractors	L.A. S.U.	Work by L.A./Statutory Undertaker i.e. statutory obligations
7	Levels and setting out of the Works	Arch. Contr.	Architect to provide information, Contractor to execute
8	Materials, good and workmanship to conform to description, testing and inspection		
8.1	Kinds and standards	Q.S. Contr. Arch.	As in B.Q.
8.2	Vouchers - materials and goods	Contr. Arch.	Of compliance - Contractor to provide if Architect requests
8.3	Inspection - tests	Arch. Contr. (Q.S.)	A.I. - cost to Employer if O.K.; Contractor if defective
8.4	Removal from the site - work materials or goods	Arch. Contr. (Q.S.)	A.I. - items not as contract
8.5	Exclusion from the Works of persons employed thereon	Arch. Contr.	A.I. transfer, not sacking
9	Royalties and Patent Rights		
9.1	Treatment of royalties etc. - indemnity to Employer	Contr. Empl. (Q.S.)	By Contractor
9.2	Architect's instructions - treatment of royalties etc.	Arch. Contr. Empl. Q.S.	Add to contract sum
10	Person-in-charge	Contr.	Always on site - Contractor's Agent

REF. NO.	SUBJECT	PERSONS PRIMARILY CONCERNED	MAJOR CONTENTS/ COMMENTS
11	Access for Architect to the Works	Contr.	Incl. Q.S. etc. - working hours
12	Clerk of Works	Empl. Arch. CoW.	Appointed by Employer, briefed by Arch. issue written directions.
13	Variations and provisional sums		
13.1	Definition of Variation	All	'All embracing' - limited by size
13.2	Instructions requiring a Variation	Arch. Contr. (Q.S.)	A.I. or confirmation
13.3	Instructions on provisional sums	Arch. Contr. Q.S.	A.I. required
13.4	Valuation of Variations and provisional sum work	Q.S. (All)	By Q.S. only, Contractor present
13.5	Valuation rules	Q.S.	As B.Q., pro-rata, fair, daywork; Claim?
13.6	Contractor's right to be present at measurement	Q.S. Contr.	
13.7	Valuations - addition to or deduction from Contract Sum		
14	Contract Sum		
14.1	Quality and quantity of work included in Contract Sum	All	B.Q.
14.2	Contract Sum - only adjusted under the Conditions - errors in computation	Empl. Contr. Q.S.	Arithmetic errors - deemed accepted by parties
15	Value added tax - supplemental provisions	Customs & Excise	
15.1	Definitions - VAT Agreement		Finance Act 1972
15.2	Contract Sum - exclusive of VAT	All	
15.3	Possible exemption from VAT		Input tax adjustment
16	Materials and goods unfixed or off-site		Transfer of title provisions
16.1	Unfixed materials and goods - on site	All	Remove only for incorporation or A.I. required
16.2	Unfixed materials and goods - off "	All	If included in a Certificate - Contractor's responsibilities pass on to supplier etc.
17	Practical completion and Defects Liability		Complete for practical purposes, NOT 'substantial completion'
17.1	Certificate of Practical Compl-etion	Arch. Empl. Contr. Q.S.	Date of Practical Completion as Certificate
17.2	Defects, shrinkages or other faults	Arch. Contr.	Schedule - make good by Contractor at his cost

REF. NO.	SUBJECT	PERSONS PRIMARILY CONCERNED	MAJOR CONTENTS/ COMMENTS
17.3	Defects etc. - Architect's instructions	Arch. Contr.	A.I. to make good prior to schedule
17.4	Cert. of Completion of Making Good Defects	Arch. Contr. (Q.S. Empl.)	Architect's opinion
17.5	Damage by frost	Contr.	Contractor's responsibility - frost prior to Practical Compl.
18	Partial possession by Employer		
18.1	Employer's wish - Contractor's consent	Empl. Arch. Contr.	Architect to certify parts and date of possession by Employer
18.1.1	Practical Completion - relevant	All	As total contract provisions but to the part
18.1.2	Defects etc. - relevant part	Arch. Contr.	" " " " " " "
18.1.3	Insurance - relevant part	Empl. Contr.	Employer's responsibility
18.1.4	Liquidated damages - relevant part	Empl. Contr. Q.S.	Pro-rata adjustment
19	Assignment and Sub-Contracts		
19.1	Assignment	Empl. Contr.	Not without consent of other party (written)
19.2	Sub-letting - Domestic Sub-Contractors - Architect's consent	Arch. Contr.	
19.3	Sub-letting - list in Contract Bills	Arch. Contr. Q.S.	Min. 3 persons
19.4	Sub-letting - determination of employment of Domestic Sub-contractor	Contr. DS/C	Immediately main Contractor's employment determined Property in materials on site, placed thereon by Domestic Sub-Contractors
19.5	Nominated Sub-Contractors	Contr. NS/C Arch. Q.S.	Part 2, nomination must occur, Contractor not obliged to execute the work etc.
20	Injury to persons and property and Employer's indemnity		
20.1	Liability to Contractor - personal injury or death - indemnity to Employer	Contr. All	Appendix-amount, unless Employer's (etc.) negligence etc.
20.2	Liability of Contractor - injury or damage to property - indemnity to Employer	Contr. All	" " due to Contractor's (etc.) negligence etc.
20.3	Injury or damage to property - exclusion of the Works and Site Materials	Contr. All	

REF. NO.	SUBJECT	PERSONS PRIMARILY CONCERNED	MAJOR CONTENTS/ COMMENTS
21	Insurance against injury to persons or property	All	
21.1	Insurance - personal injury or death - injury or damage to property	Contr. Arch.	Minimum cover as Appendix. Evidence of insurance - inspection by Employer via Architect
21.2	Insurance - liability etc. of Employer	Empl. Contr. Arch.	Appendix - whether cover required - Arch. to instruct Joint Names Policy - Indemnity amount as per Appendix note exceptions
21.3	Excepted risks	Empl.	Assumed by Employer
22	Insurance of the Works	All	
22.1	Insurance of the Works - alternative clauses	Empl. Contr. Arch. Q.S.	22A or 22B or 22C Joint Names Policy for All Risks Insurance 22A, 22B Joint Names Policy re: Specified Perils - 22C
22.2	Definitions	All	All Risks Insurance, Site Materials Special Provisions for projects in Northern Ireland
22.3	Nominated and Domestic Sub-Contractors - benefit of Joint Names Policies - Specified Perils	All	NS/Cs & DS/Cs insured under Joint Names Policy - no subrogation
22A	Erection of new buildings - All Risks Insurance of the Works by the Contractor		
22A.1	New buildings - Contractor to take out and maintain a Joint Names Policy for All Risks Insurance	Empl. Contr. Arch. Q.S.	Note end date for cover Adequacy of cover - full reinstatement value Not for Contrs. plant, huts, etc.
22A.2	Single policy - insurers approved by Employer - failure by Contractor to insure	Empl. Contr. Arch. Q.S.	Empl. to approve insurers Contr. defaults; Employer may effect insurance
22A.3	Use of annual policy maintained by Contractor - alternative to use of Clause 22A.2	Contr.	'All risks' policy; check scope of cover - Clause 22.2
22A.4	Loss or damage to Works - insurance claims - Contractor's obligations - use of insurance monies	All	Contr. - give written notice of loss/damage Contr. to repair etc. Insurance monies paid to Empl.; only these available to pay Contr. for repairs etc.

REF. NO.	SUBJECT	PERSONS PRIMARILY CONCERNED	MAJOR CONTENTS/ COMMENTS
22B	Erection of new buildings All Risks Insurance of the Works by the Employer		
22B.1	New buildings - Employer to take out and maintain a Joint Names Policy for All Risks Insurance	Empl. Contr. Arch.	New building
22B.2	Failure of Employer to insure-rights of Contractor	Empl. Contr. Arch. Q.S.	Employer to show evidence of insurance Contractor may insure and charge if Employer defaults.
22B.3	Loss or damage to Works - insurance claims - Contractor's obligations - payment by Employer	Empl. Contr. Arch. Q.S.	As Clause 22A.3 but repairs etc. paid for as a Variation under Clause 13.2
22C	Insurance of existing structures - Insurance of Works in or extensions to existing structures		
22C.1	Existing structures and contents - Specified Perils - Employer to take out and maintain Joint Names Policy	Empl. Contr. Arch. Q.S.	Insure for full cost of reinstatement etc. All monies paid to Employer
22C.2	Works in or extensions to existing structures - All Risks Insurance - Employer to take out and maintain Joint Names Policy	Empl. Contr. Arch. Q.S.	
22C.3	Failure of Employer to insure - rights of Contractor	Empl. Contr. Arch. Q.S.	Employer to produce evidence of insurance Contractor may insure and charge if Employer defaults
22C.4	Loss or damage to Works - Insurance claims - Contractor's obligations - payment by Employer	All	Contractor - written notice of loss/damage to Employer Insurance monies paid to Employer Contractor to repair etc? If so paid as Variation
22D	Insurance for Employer's loss of liquidated damages - Clause 25.4.3	Empl. Contr. Arch. Q.S.	Optional - see Appendix. If required Architect to inform Contractor to obtain quotation. A.I. re Employer's acceptance. Period as Appendix or extension of time

REF. NO.	SUBJECT	PERSONS PRIMARILY CONCERNED	MAJOR CONTENTS/ COMMENTS
23	Date of Possession, completion and postponement	All	
23.1	Date of Possession - progress to Completion Date	Contr. Arch. Empl.	Appendix. Completion by Completion Date
23.2	Architect's instructions - postponement	Arch. Contr.	Of any work
23.3	Possession by Contractor - use of occupation by Employer	All	Contractor's consent required. Insurance cover to be unaffected - consult insurers
24	Damages for non-completion		Appendix - genuine pre-estimate of Employer's loss
24.1	Certificate of Architect	Arch. Contr. Q.S.	Required as pre-requisite to payment
24.2	Payment or allowance of liquidated damages	Arch. Q.S. Contr. Empl.	Extended completion date? Any partial possession?
25	Extension of Time		
25.1	Interpretation of delay etc.	All	
25.2	Notice by Contractor of delay to progress	Contr. Arch.	Pre-requisite for each delay - Contractor to write to Architect, ? period
25.3	Fixing Completion Date	Arch. (Contr. Q.S.)	Architect - never earlier than original date in Contract
25.4	Relevant Events	Contr. Arch.	List, Contractor to specify, Architect to evaluate - 12 weeks
26	Loss and expense caused by matters materially affecting regular progress of the Works		Architect (or Q.S.) to evaluate
26.1	Matters materially affecting regular progress of the Works - direct loss and/or expense	Contr. Arch. Q.S.	Contractor's written application - as soon as apparent
26.2	List of matters	Contr. (Arch. Q.S. Empl.	Less than Cl.25; Employer's responsibility and control
26.3	Relevance of certain extensions of Completion Date	Arch.	Also entitled to claim
26.4	Nominated Sub-Contractors - matters materially affecting regular progress of the Sub- Contract Works - direct loss and/or expense	NS/C, Contr. Arch. Q.S.	

REF. NO.	SUBJECT	PERSONS PRIMARILY CONCERNED	MAJOR CONTENTS/ COMMENTS
26.5	Amounts ascertained - added to Contract Sum	All	
26.6	Reservation of rights and remedies of Contractor	Contr.	'Without prejudice'
27	Determination by Employer		Of Contractor's employment
27.1	Default by Contractor	Contr. Arch. Empl.	Grounds for determination
27.2	Contractor becoming bankrupt etc.	Contr. Empl. liquidator	Reconstruction etc.
27.3	(Number not used)		L.A. - Corruption
27.4	Determination of employment of Contractor - rights and duties of Employer and Contractor	Empl. Contr. Arch. Q.S.	Another Contractor to complete - use of plant etc. Accounts - Employer's costs
28	Determination by Contractor		
28.1	Acts etc. giving ground for determination of employment by Contractor	Contr. Empl.	
28.2	Determination of employment by Contractor - rights and duties of Employer and Contractor	Empl. Contr. Arch. Q.S.	
29	Works by Employer or persons employed or engaged by Employer		'Artists and Tradesmen'
29.1	Information in Contract Bills	Q.S. Contr. Empl.	Not part of Contract
29.2	Information not in Contract Bills	Contr. Empl.	
30	Certificates and payments		
30.1	Interim Certificates and valuations	Arch. Q.S. Contr.	Monthly (usually). 14 days to honour. By Architect.
30.2	Ascertainment of amounts due in Interim Certificates	Arch. Q.S.	Gross
30.3	Off-site materials or goods	Arch. Contr. Q.S.	Include if comply with requirements - Architect's discretion
30.4	Retention - rules for ascertainment	Arch. Q.S. Contr.	5% or 3% usually
30.5	Rules on treatment of Retention	Arch. Q.S. Contr.	
30.6.1	Final adjustment of Contract Sum - documents from Contractor - final Valuation under Clause 13	Arch. Q.S. Contr.	

REF. NO.	SUBJECT	PERSONS PRIMARILY CONCERNED	MAJOR CONTENTS/ COMMENTS
30.6.2	Items included in adjustment of Contract Sum	Arch. Q.S. Contr.	
30.6.3	Computation of adjusted Contract Sum - Contractor to receive copy		
30.7	Interim Certificate - final adjustment or ascertainment of Nominated Sub-Contract Sums	NS/C, Arch. Q.S.	
30.8	Issue of Final Certificate	Arch. Q.S. Contr.	3 months from end D.L.P., Completion of m.g. defects etc.
30.9	Effect of Final Certificate	All	Evidence of compliance
30.10	Effect of certificates other than Final Certificate	All	
31	Finance (No.2) Act 1975 - statutory tax deduction scheme	Q.S.	
31.1	Definitions		
31.2	Whether Employer a 'contractor'	Empl. Contr.	
31.3	Provision of evidence - tax certificate	Contr.	Type of certificate - inspection etc.
31.4	Uncertified Contractor obtains tax certificate - expiry of tax certificate - cancellation of tax certificate	Empl. Contr.	
31.5	Vouchers	Contr.	Essential - payments liability etc.
31.6	Statutory deduction - direct cost of materials	Contr. Empl.	Applies to labour content only
31.7	Correction of errors	Empl. Contr.	
31.8	Relation to other clauses		
31.9	Application of arbitration agreement	Empl. Arch. Contr. Arbitrator	
32	Outbreak of hostilities		
32.1	Notice of determination of the Contractor's employment	Govt. Empl.	General mobilisation of UK armed forces
32.2	Protective work etc.	Arch. Contr.	A.I.
32.3	Payment	Arch. Q.S. Contr.	

REF. NO.	SUBJECT	PERSONS PRIMARILY CONCERNED	MAJOR CONTENTS/ COMMENTS
33	War damage		
33.1	Effect of war damage	Arch. Contr.	
33.2	Relation with Clause 32	Arch.	
33.3	Use of compensation for war damage	All	
33.4	Definition of war damage		
34	Antiquities	Arch. CoW	
34.1	Effect of find of antiquities	Contr.	Stop adjacent work. Contractor to inform Architect or CoW
34.2	Architect's instructions on antiquities found	Arch.	
34.3	Direct loss and/or expense	Arch. Q.S. Contr.	Architect (or Q.S,) to ascertain

CONDITIONS: PART 2: NOMINATED SUB-CONTRACTORS AND NOMINATED SUPPLIERS

REF. NO.	SUBJECT	PERSONS PRIMARILY CONCERNED	MAJOR CONTENTS/ COMMENTS
35	Nominated Sub-Contractors General	NS/C	
35.1	Definition of a Nominated Sub-Contractor	Q.S. Contr.	P.C. Sum in B.Q. or by A.I. re provisional sum
35.2	Contractor's tender for works otherwise reserved for a Nominated Sub-Contractor	Contr.	Contractor may be a NS/C
35.3	Documents relating to Nominated Sub-Contractors Procedure for nomination of a Sub-Contractor	All	NSC/1,2,3 & 4 or NSC/2a and 4a
35.4	Contractor's right of reasonable objection to proposed sub-contractor	Contr.	a.s.a.p. - within 7 days of A.I.
35.5	Use of Tender NSC/1 - circumstances where Tender NSC/1 is not used	Arch. Contr.	
35.6	Limit on nomination	All	35.11. and 35.12. or only if NSC/1 & 2 used
35.7	Architect's preliminary action prior to nomination of a proposed sub-contractor - duty of Contractor	Arch. Contr. NS/C	Architect send NSC/1, 2 and notice of nomination to Contr.
35.8	Contractor and proposed sub-contractor - failure to agree - Contractor's and Architect's duty	Contr. Arch. NS/C	10 working days Contractor - inform Architect and reasons Architect - issue A.I.
35.9	Proposed sub-contractor - withdrawal of offer	NS/C. Contr. Arch.	Contractor to write to Architect and no further action

REF. NO.	SUBJECT	PERSONS PRIMARILY CONCERNED	MAJOR CONTENTS/ COMMENTS
35.10	Receipt of completed Tender - issue of nomination instruction by Architect	Arch.	NSC/1 completed NSC/3 issued
35.11	Tender NSC/1 and Agreement NSC/2 not issued - use of Agreement NSC/2a	Arch.	A.I. Advise use of NSC/1a and NSC/3a
35.12	Sub-Contract NSC/4a	All	
35.13	Payment of Nominated Sub-Contractor		
35.13.3 & 2	Architect - direction as to interim payment for Nominated Sub-Contractor	Arch. NS/C. Contr. (Empl.) Q.S.	In Interim Certificate Architect - inform NS/Cs
35.13.3 to 5	Direct payment of Nominated Sub-Contractor	Arch. NS/C. Contr. Empl. Q.S.	If Contractor defaults, Architect to certify
35.14.1 & 2	Extension of period or periods for completion of Nominated Sub-Contract Work	NS/C. Contr. Arch.	Contractor grants with Architect's written consent
35.15.1 & 2	Failure to complete Nominated Sub-Contract Works	Contr. Arch.	Architect to certify in writing to Contractor - within 2 months of Contractor's written notice of NS/C's failure
35.16	Practical Completion of Nominated Sub-Contract Works	Arch.	Architect to certify
35.17 to 19	Final Payment of Nominated Sub-Contractors	Arch. Empl. Q.S.	Within 12 months from 35.16 certificate
35.18	Defects in Nominated Sub-Contract Works after final payment of Nominated Sub-Contractor - before issue of the Final Cert.	NS/C. Arch. Contr.	Failure to remedy - set-off
35.19	Final payment - saving provisions	Arch. Q.S. Contr. Empl.	
35.20	Position of Employer in relation to Nominated Sub-Contractor	NS/C. Empl.	Only as NSC/2 or 2a
35.21	Clause 2 of Agreement NSC/2 or Clause 1 of Agreement NSC/2a - position of Contractor	Contr.	
35.22	Restrictions in Contracts of Sale etc. - Limitation of liability of Nominated Sub-Contractors	NS/C, Contr. Empl.	To be passed on to Employer

REF. NO.	SUBJECT	PERSONS PRIMARILY CONCERNED	MAJOR CONTENTS/ COMMENTS
35.23	Position where Proposed nomination does not proceed further	Arch.	A.I.
35.24	Circumstances where re-nomination necessary	NS/C. Contr. Arch.	Default, liquidation etc. of NS/C
35.25 & 35.26	Determination of Employment of Nominated Sub-Contractor - Architect's instructions	Arch. NS/C. Contr. Q.S.	A.I. required
36 NOMINATED SUPPLIERS		N Sup.	
36.1	Definition of Nominated Supplier	Arch. Q.S.	P.C. sum or single source
36.2	Architect's instructions	Arch. Contr. Q.S.	To Nominate
36.3	Ascertainment of costs to be set against prime cost sum	Arch. Q.S. Contr. N Sup.	Architect's opinion
36.4	Sale contract provisions - Architect's right to nominate supplier	Arch. Contr.	
36.5	Contract of sale - limitation or exclusion of liability	Arch. Contr. N Sup. Empl.	Architect's written approval - liability of Contractor to Employer so limited
CONDITIONS: PART 3: FLUCTUATIONS			
37	Choice of fluctuation provisions - entry in Appendix	Empl. Q.S. Arch.	
Note: Clauses 38, 39 and 40 are published separately.			
38 CONTRIBUTION, LEVY AND TAX FLUCTUATIONS			
38.1.1	Deemed calculation of Contract Sum - types and rates of contribution etc.		
38.1.2	Increases or decreases in rates of contribution etc. - payment or allowance		

REF. NO.	SUBJECT	PERSONS PRIMARILY CONCERNED	MAJOR CONTENTS/ COMMENTS
38.1.3 & 4	Persons employed on site other than 'workpeople'		Craft operative rate fluctuations applicable
38.1.5 to 7	Refunds and premiums		
38.1.8	Contracted-out employment		Deemed not contracted out
38.1.9	Meaning of contribution etc.		
38.2.1 & 2	Materials - duties and taxes		
38.3	Fluctuations - work sub-let - Domestic Sub-Contractors		
38.4 to 6	Provisions relating to Clause 38		
38.4.1	Written notice by Contractor		
38.4.2	Timing and effect of written notices		Reasonable time of event; condition precedent to recovery
38.4.3	Agreement - Quantity Surveyor and Contractor		They may agree
38.4.4	Fluctuations added to or deducted from Contract Sum		Not subject to Retention
38.4.5	Evidence and computations by Contractor		To Architect (or Q.S.)
38.4.7	No alteration to Contractor's profit position where Contractor in default over completion		This paid net No fluctuations for events after 'scheduled Completion Date'
38.5	Work etc. to which clauses 38.1 to 3 are not applicable		Daywork etc.
38.6	Definitions for use with Clause 38		
38.7	Percentage addition to fluctuation payments or allowances		
39 LABOUR AND MATERIALS COST AND TAX FLUCTUATIONS			
39.1.1	Deemed calculation of Contract Sum - rates of wages etc.		Applicable or promulgated at Date of Tender
39.1.2	Increases or decreases in rates of wages etc. - payment or allowance		

REF. NO.	SUBJECT	PERSONS PRIMARILY CONCERNED	MAJOR CONTENTS/ COMMENTS
39.1.3 & 4	Persons employed on site other than workpeople		As Craft operative
39.1.5 & 6	Workpeople - wage-fixing body - reimbursement of fares etc.		
39.2.1 to 8	Contributions, levies and taxies		Note CITB
39.3.1 to 3	Materials, goods, electricity and fuels		Basic list, market price
39.4	Fluctuations - work sub-let - Domestic Sub-Contractors		
39.4.1	Sub-let work - incorporation of provisions to like effect		
39.4.2	Sub-let work - fluctuations - payment to or allowance by the Contractor		
39.5 to 7	Provisons relating to Clause 39		
39.5.1	Written notice by Contractor		A.S.A.P. to Architect (or Q.S.)
39.5.2	Timing and effect of written notices		Condition precedent to Contractor's recovery
39.5.3	Agreement - Quantity Surveyor and Contractor		May agree
39.5.4	Fluctuations added to or deducted from Contract Sum		
39.5.5	Evidence and computations by Contractor		
39.5.6	No alteration to Contractor's profit		Paid net
39.5.7	Position where Contractor in default over completion		Not in respect of events after Completion Date
39.6	Work etc. to which clauses 39.1 to 4 not applicable		Daywork etc.
39.7	Definitions for use with Clause 39		
39.8	Percentage addition to fluctuation payments or allowances		
40	USE OF PRICE ADJUSTMENT FORMULAE		
40.1	Adjustment of Contract Sum - price adjustment formulae for building contracts - Formula Rules		

REF. NO.	SUBJECT	PERSONS PRIMARILY CONCERNED	MAJOR CONTENTS/ COMMENTS
40.2	Amendment to Clause 30 - interim valuations and payments		Interim Valuation required for each Interim Certificate
40.3	Fluctuations - articles manufactured outside the United Kingdom		Market Price at Date of Tender - use 'manual' system
40.4.1 40.4.2	Nominated Sub-Contractors Domestic Sub-Contractors		Appropriate specialist formulae or 'some other method'
40.5	Power to agree - Quantity Surveyor and Contractor		May agree alterations to the methods and procedures
40.6	Position where Monthly Bulletins are delayed, etc.		'Fair' adjustments and use indices retrospectively when available
40.7	Formula adjustment - failure to complete		Over-run - use indux at Completion Date

REF. NO.	SUBJECT	PERSONS PRIMARILY CONCERNED	MAJOR CONTENTS/ COMMENTS
APPENDIX			
SUPPLEMENTAL PROVISIONS (the VAT Agreement)		Empl. Contr.	
1	Interim payments - addition of VAT		
1.1	Written assessment by Contractor		Amounts subject to positive rate of tax and rate(s) applicable
1.2	Employer to calculate amount of tax due - Employer's right of reasonable objection		
1.3	Written final statement - VAT liability of Contractor - recovery from Employer		
1.4	Contractor to issue receipt as tax invoice		When Contractor receives payment
2	Value of supply - liquidated damages to be disregarded		
3	Employer's right to challenge tax claimed by Contractor	Commissioners	Refer to Commissioners?
4	Discharge of Employer from liability to pay tax to the Contractor		
5	Awards by Arbitrator or court		
6	Arbitration provision excluded		
7	Employer's right where receipt not provided		

Bibliography/Sources

Hudson's Building and Engineering Contracts (10th Edition) and Supplement, I.N. Duncan Wallace, Sweet and Maxwell.

Contractors' Guide to the Joint Contracts Tribunal's Standard Forms of Building Contract, 1978, Vincent Powell-Smith, IPC Building and Contract Journals Ltd.

The Standard Forms of Building Contract, Walker-Smith and Close, Charles Knight & Co. Ltd.

J.C.T. Guide to the Standard Form of Building Contract 1980 Edition, Joint Contracts Tribunal, RIBA Publications Ltd.

Introduction to English Law (10th Edition), P.S. James, Butterworths.

John Sims - series of contributions concerning the J.C.T. Standard Form of Building Contract 1980 Edition - from Building, commencing 5 February 1980.

Determination of Employment under the Standard Forms of Contract for Construction Works, A. T. Ginnings, The Quantity Surveyor, February 1978, pp 97-101.

Building Law Reports, J. Parris (ed), George Goodwin.

Construction Law Reports, V. Powell-Smith, Architectural Press.

V. Powell-Smith - regular articles in Contract Journal.

SUPPLEMENT

Application: JCT 80 - Private with Quantities - in full
JCT 80 - other editions - with some qualifications

Guidance on how the amendment should be implemented is given on page 2 of the amendment - either the Contract may be altered and the revised Appendix substituted for the original Appendix, alterations must be initialled by the parties; or the amendment may be attached to the original Contract form, again the amendment and the new Appendix must be initialled by the parties.

Amendments:

Article 3 Architect and
Article 4 Quantity Surveyor

Employer must re-nominate within a reasonable time, not exceeding 21 days (of death or cessation).
Contractor has maximum of 7 days to object to any re-nomination.

Article 5 Settlement of disputes - Arbitration

New Article inserted.
Arbitration is retained as the main means of settling disputes but now forms Part 4 of the Contract, Clause 41.

Clause 1.3 Definitions

'Date of tender' is replaced by 'Base Date', the date of the latter is stated in the Appendix. (Many consequential adjustments of terminology in the Contract.) Helpful for Formulae Fluctuations.

Clause 2.3 **Discrepancies in or divergencies between documents**

Clause 2.3.5 the Numbered Documents

Clause 1.3 Numbered Documents - 'any document referred to in the first recital in any sub-contract with a Nominated Sub-contractor.'

Clause 26.2.3 inclusion of the Numbered Documents to provide possible grounds for a loss and expense claim.

Clause 7 **Levels and setting out of the Works.**
Contractor responsible for the Contractor's errors in setting out and for their amendment at no cost to the Employer. Provided the Employer so agrees, the Architect may instruct the Contractor not to amend any such errors in setting out. Appropriate deduction from the Contract Sum to be made for such errors not amended.

Clause 8.1 **Kinds and Standards**
So far as procurable, all materials and goods must be of the kinds and standards described in the Contract Bills. However, materials and goods must be to the reasonable satisfaction of the Architect where so required under Clause 2.1.

Thus, it appears, that if the stipulated materials, etc. are not available, a substitute, acceptable to the Architect, must be provided.

Clause 8.1.2 Workmanship to accord with requirements given in BQ but, if not so specified, to be of standards appropriate to the Works and to the reasonable satisfaction of the Architect where so require under Clause 2.1.

Clause 11 **Access for Architect to the Works**
Access may be retricted reasonably to protect any proprietary rights of Contractors, Dom S/C or NS/C.

Clause 13.1.2 **Definition of Variation**
Impositions or changes of restrictions, etc. covered by the sub clauses 1 to 4 or imposed by the Employer in the BQ are now included.

Clause 13.2 Instructions requiring a Variation
Power of the Architect to instruct a Variation is subject to the Contractor's right of reasonable objection as per Clause 4.1.1.

Clause 13.4.1 Valuation of Variations and provisional sum work
This Clause is re-numbered.
Provisions of first paragraph now subject to limitation imposed by second paragraph - valuation of NS/C items is as per the NSC.

Clause 13.5.5 Valuation rules
All AIs regarding Variations or expenditure of a provisional sum are now covered by this Clause.

Clause 17.2 Defects, shrinkages or other faults
Defects, etc. to be remedied by the Contractor at no cost to the Employer. However, if the Architect, with the Employer's consent, instructs the Contractor to do otherwise, an appropriate deduction must be made from the Contract Sum in respect of such defects, etc.

Clause 17.3 Defects, etc. - Architect's instructions
Final part deleted to give effect to amendment to Clause 17.2.

Clause 19.1 Assignment
Existing Clause re-numbered 19.1.1.
Clause 19.1.2: Provided the Appendix states that Clause 19.1.2 applies, if the Employer transfers his/her interest in the property (whole or part) to a third party, the Employer may also, prior to Practical Completion of the Works, assign the Employer's rights to bring proceedings (arbitration or litigation) to enforce the terms of the Contract to that third party (also covers Employer's granting a leasehold interest in the property to a third party). Third party is estopped from disputing any enforceable agreements between the Employer and Contractor arising out of the Contract provided the agreements were made before the date of the assignment.

Appendix: New entry to note whether Clause 19.1.2 applies or not.

Clause 19.5 Nominated Sub-Contractors
Clause 19.1 Subject to agreement, the Contractor may be required to execute items which were to be executed by a NS/C.

Clause 21.1.1.1 Contractor's insurance - personal injury or death - injury or damage to property.
References to S/Cs deleted

21.1.1.2)
21.1.2) References to S/Cs deleted
21.1.3)

The insurance responsibility is now the Contractor's; any insurance which may be effected by S/Cs is a matter to be agreed between the Contractor and the S/Cs.
The Employer is protected by the Contractor's indemnity under Clause 20.

Clause 23.1 Date of possession —progress to Completion Date.
Existing Clause 23.1 re-numbered 23.1.1

New Clause 23.1.2: Employer may defer giving possession of the site for a period as stated in the Appendix but not exceeding 6 weeks from the Date of Possession.

New Clause 25.4.13: Application of deferment of possession under Clause 23.1.2 is a Relevant Event.
Clause 26.1 amended to allow direct loss and/or expense to be claimed for deferment of possession under Clause 23.1.2.

Appendix: New entries to state whether Clause 23.1.2 applies or not and, if it does apply, the period of deferment (maximum) to be specified if it is to be less than 6 weeks.

The inclusion of the deferment provision allows the Employer to delay giving possession of the site for up to the stated maximum period without being in breach of contract (and, hence, time becoming at large).

Clause 25.3.1 Fixing Completion Date
New paragraph inserted at the end of the existing clause allowing the Architect to not fix a new Completion Date in the light of the information provided by the Contractor. The Architect must inform the Contractor in writing within the usual 12 week period or not later than the Completion Date (if less than 12 weeks). Thus, in respect of each application by the Contractor for an extension of time, the Architect must decide and inform the Contractor, in writing, of the decision and any new Completion Date.

Clause 25.3.2 Fixing Completion Date
Omissions of obligations or restrictions now provide grounds for the Architect to revise the Completion Date to an earlier one (but still no earlier than the original Completion Date).

Clause 25.3.3.2 now includes omissions of obligations or restrictions; following Clause 25.3.2.

Clause 25.3.3 Fixing Completion Date
The requirement for the Architect to revise or confirm the Completion Date in writing to the Contractor is either:

a) obligatory within 12 weeks of Practical Completion, but
b) discretionary (subject to (a)) after the Completion Date if that date is prior to the date of Practical Completion.

Thus the Completion Date may be fixed finally prior to the Contractor's achieving Practical Completion.

Clause 26.4.1 Nominated Sub-contractors - matters materially affecting regular progress of the Sub-Contract Works - direct loss and/or expense.
Includes deferment of giving possession of the site where Clause 23.1.2 applies.

Clause 27.1.3 Determination by Employer - default by Contractor
'persistently' is deleted. Any non-compliance with a notice is now sufficient ground.

Clause 27.2 Contractor becoming bankrupt, etc.
Changes made to encompass the provisions of the Insolvency Act, 1986.

Clause 27.3 Corruption (new Clause)
A widely-scoped clause, now applicable to the Private edition as well as the Local Authorities' edition.

Corruption by/on behalf of the Contractor (with or without the Contractor's knowledge) under this or any other contract with the Employer or if any Contractor's employee or any person acting on behalf of the Contractor has committed an offence under the Prevention of Corruptions Acts 1889 to 1916 in relation to such contracts, the Employer is entitled to determine the employment of the Contractor under this or any other contract.

This is the only provision where acts under other contracts between the parties affect the particular contract (and vice-versa).

Clause 28.1.3 Determination by Contractor - Acts, etc. giving ground for determination of employment by Contractor.

Clause 28.1.2.1 force majeure)
Clause 28.1.3.2 loss/damage caused by specified perils) deleted
Clause 28.1.3.3 civil commotion)

Clauses 28.1.3.4 to 28.1.3.7 renumbered as Clauses 28.1.3.1 to 28.1.3.4.

Clause 28.1.3.5 repeat of Clause 25.4.12 as ground for determination of employment by the Contractor.

Clause 28.1.4 changes made to encompass the provisions of the Insolvency Act, 1986.

Clause 28.1.3.1 Determination by Contractor - Acts etc.
(was Clause 28.1.3.4) **giving ground for determination of employment by Contractor**

The clause now excludes negligence or defaults by the Contractor's servants or agents and persons employed or engaged in connection with the Works (whole or part) including such person's servants or agents other than:

NS/Cs. (see J Jarvis Ltd v Rockdale Housing Association Ltd (1987))

Employer

People engaged (etc.) by the Employer or by any Local Authority or statutory undertaker executing work solely in pursuance of their statutory obligations.

Note For exemption from this provision, the work executed must be performed under a statutory obligation and not merely incidental to any such obligation (e.g. excavation for a power cable by the electricity authority would not be excepted as it is not carried out under a statutory obligation).

Appendix: Period of delay under Clause 28.1.3 to be inserted (recommended one month).

Clause 28A Determination by Employer or Contractor
(new Clause) (Replaces Clauses previously numbered 28.1.3.1, 2 and 3).

.1 If the execution of the whole or the majority of the uncompleted works, except making good defects etc. under Clause 17, is suspended for a continuous period as stated in the Appendix, without prejudice to other rights or remedies, the Employer or the Contractor may determine the employment of the Contractor under this Contract. The determination is effected by either party giving the other written notice (registered post or recorded delivery); the notice must not be given unreasonably or vexatiously.

The expressed reasons for the notice of determination and the suggested (footnote to Appendix) lengths of time applicable are:

.1 force majeure - one month, or

.2 loss or damage to the Works caused by any one or more of the Specified Perils - three months, or

.3 civil commotion - one month.

The periods of delay which are to apply must be inserted in the Appendix.

.2 The Contractor cannot give notice under Clause 28A.1 where the loss or damage to the Works due to the occurrence of the Specified Perils was caused by negligence or default of the Contractor or those for whom the Contractor is responsible.

.3 Once determination of the Contractor's employment under Clause 28A.1 has occurred, except for Clause 28.2.2.6 (direct loss, damage caused to Contractor or any NSC by the determination), settlement of the contract accounts must be executed under Clause 28.2.

Clause 30.2 Ascertainment of amounts due in Interim Certificates

Clause 30.2.1.1 excludes the prices for restoration, repair of loss or damage and removal and disposal of debris (Clauses 22B.3.5 and 22C.4.4.2) from deduction of Retention; the prices for such work are calculated under the provisions for valuing Variations and so should be distinguished from ordinary Variations.

Clause 30.2.2.1 - delete 22B and 22C, replace with 22B.2 and 22C, - any premiums paid by the Contractor are added to the Contract Sum - to be included in Interim Certificates and not subject to Retention.

Clause 30.2.2.2 includes prices of restoration work, etc. under Clauses 22B.3.5 and 22C.4.4.2, valued as Variations, not subject to Retention.

Clause 30.6.2 Items included in adjustment of Contract Sum.

Clause 30.6.2.1 includes as a deduction from the Contract Sum, the certified value of any work by a NS/C, whose employment has been determined under Clause 35.24, which was not in accordance with the relevant sub-contract but which has been paid or otherwise discharged by the Employer.
See Fairclough Building Ltd v. Rhuddlan Borough Council (1985): The revised Clause 30.6.2 enables the Employer to obtain credit from the Contractor for a NS/C's work which was not in accordance with the NSC, whose employment has been determined under Clause 35.24 and for which the Employer has made payment in some form (usually payment of the NS/C via the Contractor).

Clauses 30.6.1 and 30.8 Final adjustment of Contract Sum - documents from Contractor - final valuations under Clause 13 and Issue of Final Certificate.

Clause 30.6.1.1 The Contractor must supply the documents necessary for the adjustment of the Contract Sum to the Architect or, if so instructed, to the QS within 6 months of the Practical Completion of the Works.

Clause 30.6.1.2 existing Clause deleted and replaced by:-
Clause 30.6.1.2 Within 3 months of receipt (by Architect or QS) of the documents noted in Clause 30.6.1.1 (documents for adjustment of the Contract Sum):

.1 The Architect or QS must ascertain (unless done previously) any loss/expense under Clauses 26.1, 26.4.4 and 34.3, and

.2 The QS must prepare a statement of all the adjustments to be made to the Contract Sum under Clause 30.6.2, except loss/expense amounts ascertained under Clause 30.6.1.2.1 .
The Architect must forthwith send to the Contractor, with relevant extracts to each NS/C, copies of:

any ascertainment - Clause 30.6.1.2.1, and

the statement - Clause 30.6.1.2.2.

Clause 30.8 First sentence deleted and replaced by:-
The Architect must issue the Final Certificate, and inform each NS/C of the date of its issue, within 2 months at the latest of:-

a) the end of the DLP, or

b) the date of issue of the Certificate of Completion of Making Good Defects, or

c) the date on which the Architect sent a copy of any ascertainment under Clause 30.6.1.2.1 and of the statement under Clause 30.6.1.2.2 to the Contractor.

Any balance due, from the Employer to the Contractor, or vice-versa, is a debt payable from the 28th day after the date of the Final Certificate (was after the 14th day).

Clause 30.6.3 Clause deleted

Appendix: Entry for Clause 30.6.1.2 - deleted.

Clause 30.9.3 Period amended from 14 days to 28 days - Final Certificate's role as conclusive evidence.

Clause 30.9 Effect of Final Certificate

Clause 30.9.1.1 Clarifying qualification added.
New conclusive evidence provided by the Final Certificate that:

.3 all and only such extensions of time, if any, as are due under Clause 25 have been given

.4 reimbursement of any direct loss/expense claims under Clause 26.1 is final settlement of all and any claims by the Contractor due to the occurrence of matters under Clause 26.2.

The conclusive evidential effects of the Final Certificate are subject to challenge under Clause 30.9.3.

Clause 35.13 Payment of Nominated Sub-Contractor

Clause 35.13.6 (new Clause) If under the applicable Employer - NS/C Agreement (Clause 2.2 of NSC/2 or Clause 1.2 of NSC/2a) and prior to the issue of the AI of nomination, the Employer has paid the S/C an amount for design work/materials, etc. /fabrication which is included in the Sub-Contract Sum or Tender Sum:

.1 The Employer must send to the Contractor the written statement of the NS/C of the amount to be credited to the Contractor, and

.2 the Employer may deduct from Interim Certificates which include interim or final payments to the NS/C amounts up to the credit noted under Clause 35.13.6.1; maximum deductions for such credits in any Interim Certificate are amounts directed by the Architect to be included in that Interim Certificate for the relevant NS/Cs.

Clause 30.7 The reference to credits under NSC/2 or 2a is deleted.

Note: Any direct payments from the Employer to any NS/C are ignored in calculating amounts due in either Interim Certificates or in the Final Certificate.
The Contractor may pass on any such deductions when paying the NS/C.
Any queries regarding VAT caused by the operation of this Clause should be referred to the VAT office.

Clause 35.24 Circumstances where re-nomination necessary

Clause 35.24.4 The Employer has required the Contractor to determine the employment of the NS/C under Clause 29.3 of NSC/4 or 4a and the employment of that NS/C has been determined.

Existing Clause:	Renumbered as Clause:
35.24.4	35.24.5
35.24.4.1	35.24.5.1
35.24.4.2	35.24.5.2
35.24.4.3	35.24.5.3*
35.24.5	35.24.6*
35.24.6	35.24.7*

* Clauses amended also:-
35.24.5.3, 35.24.6 and 35.24.7:-

The re-nomination of a NS/C by the Architect is to include making good of work, re-supply of materials, etc. by the NS/C whose employment was determined where that work etc. was not in accordance with the Sub-Contract, ie the new nominations must include for making good defective work, materials, etc. of the original NS/C.

Clause 35.24.5.3 The Contractor shall agree (agreement not to be withheld unreasonably) the price of the new NS/C.

Clause 35.24.6 Includes instances where Clause 35.24.2 or Clause 35.24.4 apply.

Clause 35.24.7 (new numbering) Provisions regarding extra amounts payable due to re-nomination are deleted.

Clause 35.24.8 Amended to permit deduction by the Employer of any additional amounts which the Employer must pay to the new NS/C, (where Clauses 35.24.3 and 35.24.7 apply and the original NS/C determined his/her employment under the Sub-Contract validly) from sums due to the Contractor after the certification of such additional sums from monies due to or to become due to the Contractor under the Contract. Such sums may be recovered from the Contractor by the Employer as a debt.

Clause 35.24.9 After the obligation for re-nomination has occurred, the Architect must re-nominate within a reasonable time under Clauses 35.24.5.3, 35.24.6 and 35.24.7.

The circumstances will be important in deciding what constitutes a reasonable time.

The revisions to Clause 35 give effect to the Court of Appeal's decision in the case of Fairclough Building Limited v. Rhuddlan Borough Council (1985), as noted in the main text of this book.

Clause 36 Nominated Suppliers

Clause 36.4.3 A programme for delivery of the Nominated Supplier's items is to be agreed between the N.Sup. and the Contractor, or if no such programme is agreed, delivery must accord with the Contractor's reasonable directions.

The agreed delivery programme now may be varied on the following grounds:-

force majeure

civil commotion, etc.

AIs under Clause 13.2 or 13.3.

failure of Architect to supply information to the N.Sup. within due time; N.Sup. to have applied in writing for the information at a reasonable time for when the information was required.

exceptionally adverse weather.

Clause 36.4.8 Deleted and replaced by the following:-
Clause 36.4.8.1 If questions of law arise out of any dispute between
the Contractor and the N.Sup. which has been referred to arbitration, the Contractor or N.Sup. may:

a) appeal to the High Court on any question of law arising out of an award made in the arbitration, and

b) apply to the High Court to determine any question of law arising in the course of the arbitration.

The Contractor and N.Sup agree that the High Court has jurisdiction to determine such questions of law.

Clause 36.4.8.2 This Clause is an amendment of Clause 36.4.8 to accord with the revised arbitration provisions, now forming Clause 41 and the provisions of Clause 36.4.8.1.

Clause 38 Contribution, levy and tax fluctuations
Clause 39 Labour and materials cost and tax fluctuations

Clause 38.6.4. Revised definition of 'wage-fixing body' by
Clause 39.7.4 defining 'recognised terms and conditions' as concerning:-

- trade or industry of the employer

- workers in comparable employment in that trade/industry are 'governed' by the terms and conditions

- settlements by agreement or award, the parties to which are employers' associations and independent trade unions which represent substantial proportions of relevant employers and workers.

Clause 40.4 Use of price adjustment formulae

Text deleted and 'number not used' inserted.

After Clause 37 insertions are made referring to Clauses 38, 39 and 40 and to the effect that those three clasues are published separately.

(New) **Part 4: Settle of disputes - Arbitrations**

Clause 41.1 Reference to arbitration is to be made if a dispute/difference, as per Article 5, has arisen including:

a) items left to the Architect's discretion

b) The Architect's withholding any certificates to which the Contractor may claim to be entitled

c) Adjustment of the Contract Sum under Clause 30.6.2

d) Rights and duties of the parties under Clauses 27, 28 32 or 33.

e) unreasonable withholding of consent by Employer, Architect or Contractor

The Arbitrator is:

a) a person to be agreed between the parties to act as Arbitrator, or

b) someone appointed by the person named in the Appendix to act as Arbitrator.

The parties have 14 days from either party's giving a written request to the other to concur in the appointment of an Arbitrator to agree upon the Arbitrator; if no such agreement is obtained, the person named in the Appendix must appoint the Arbitrator.

In the Appendix, the person to appoint the Arbitrator is the President or Vice-President of RIBA, RICS or Chartered

Institute of Arbitrators; if no appointee is identified in the Appendix, the President or Vice-President of RIBA is the appointor.

Clause 41.2.1 Update of Article 5.1.4 to incorporate Clause 41.6 appeals/applications to the High Court on points of law arising from the arbitration/awards regarding joined references.

Clause 41.2.2 As Article 5.1.5.

Clause 41.2.3 as Article 5.1.6 but with references to clauses amended.

Clause 41.3 as Article 5.2 but including:

a) references on questions of whether determination under Clause 22C.4.3.1 will be just and equitable

b) disputes/differences regarding withholding of consent by the Contractor under Clauses 18.1 and 23.3.2

as being not referable until Practical Completion (actual or alleged) or determintion of the Contractor's employment (actual or alleged) has occurred.

Clause 41.4 as Article 5.3

Clause 41.5 That the Arbitrator's award is final and binding on the parties is now subject to Clause 41.6 - High Court's resolving points of law.

Clause 41.6 The parties agree that either may (ss 1(3)a and 21(1)(b), Arbitration Act, 1979):

1 appeal to the High Court on any question of law arising out of an award

2 apply to the High Court to determine any question of law arising during the reference.

The parties agree that the High Court has the jurisdiction to determine any such questions of law.

Clause 41.7 as Article 5.5

The Amendment notes the changes to be made to the Private with Approximate Quantities and Without Quantities and the Local Authorities' versions of the Contract.

Amongst the more notable changes is the deletion of the Supervising Officer as the alternative to the Architect and the consequent introduction of the Contract Administrator to fill the role.

Amendment to JCT 80 issued January 1988

Amendment to Clause 8.4 (applies to all versions of JCT 80)

Existing Clause 8.4 is deleted, together with the side heading, and is replaced by a much more comprehensive provision.

Clause 8.4 Powers of Architect - work not in accordance with the Contract.

If any work, materials or goods are not in accordance with the Contract, the Architect may:

1 Issue an AI regarding removal from the site of such work, materials, etc. (all or part).

2 Allow all/any of such work, materials, etc. to remain, provided:

a) the Employer so agrees,

b) the allowance is made after consultation with the Contractor (who must consult any relevant NS/C(s) immediately), and

c) the allowance is confirmed by the Architect in writing.

Such allowance is NOT a Variation under the Contract and an appropriate DEDUCTION for the allowance must be made in the adjustment of the Contract Sum.

3 After consultation with the Contractor (who must consult any relevant NS/Cs immediately), issue any necessary AIs which are consequent upon the action taken under Clauses 8.4.1 or 8.4.2; these AIs do NOT give grounds for:

a) additions to the Contract Sum under Clauses 13.4 or 26, or

b) extensions of time under Clause 25.

4 Issue AIs to open up for inspection or to test for any further similar non-compliance.
Such AIs must:

a) be reasonable in the circumstances,

b) be issued following due regard to the Code of Practice relating to Clause 8.4 (and, thus, appended to the Conditions),

c) be designed to establish, to the Architect's reasonable satisfaction, the likelihood/extent of further non-compliance.

Provided the AIs so issued are reasonable, no addition to the Contract Sum under Clauses 8.3 and 26 may be made whatever the results of the inspection/tests.

However, an extension of time may be awarded under Clause 25.4.5.2 unless the work, etc. was revealed by the inspection/tests not to be in accordance with the Contract.

In consequence of the amendment to Clause 8.4 the following alterations to other Clauses may be made:

Clause 30.2.3.1 Existing Clause deleted and replaced by - any amount deductible under:

a) Clause 7 - levels and setting out

b) Clause 8.4.2 - remaining of non-conforming work etc.

c) Clause 17.2 - defects, shrinkages, etc.

d) Clause 17.3 - defects, shrinkages, etc.

and any amount(s) allowable by the Contractor to the Employer under the 'traditional' fluctuations provisions - Clauses 38 and 39.

Clause 30.6.2.4 Includes any amount deductible or deducted under Clauses 7, 8.4.2, 17.2 or 17.3.

Clause 41.3.3 Amended to allow disputes under Clause 8.4 to be referred to Arbitration prior to Practical Completion.

Clause 41.4 The Arbitrator may not examine a decision of the Architect to issue AIs under Clause 8.4.1 - removal from the site of non-conforming work, etc.

The Code of Practice, to which reference is made in Clause 8.4, gives guidance on how an Architect should proceed and what should be considered in evaluating the possibility of further non-conforming of work, etc.